Florida's Farmworkers in the
Twenty-first Century

The Florida History and Culture Series

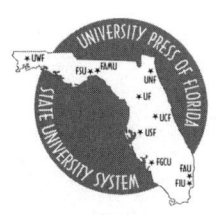

Florida A&M University, Tallahassee
Florida Atlantic University, Boca Raton
Florida Gulf Coast University, Ft. Myers
Florida International University, Miami
Florida State University, Tallahassee
University of Central Florida, Orlando
University of Florida, Gainesville
University of North Florida, Jacksonville
University of South Florida, Tampa
University of West Florida, Pensacola

Text by Nano Riley

Photographs by Davida Johns

Foreword by Raymond Arsenault and Gary R. Mormino

University Press of Florida
Gainesville Tallahassee Tampa Boca Raton Pensacola Orlando Miami Jacksonville Ft. Myers

Florida's Farmworkers in the Twenty-first Century

Copyright 2002 by Nano Riley
Photographs copyright 2002 by Davida Johns
Printed in the United States of America on acid-free paper
All rights reserved

07 06 05 04 03 02   6 5 4 3 2 1

Library of Congress Cataloging-in-Publication Data
Riley, Nano, 1942–
Florida's farmworkers in the twenty-first century / text by Nano Riley; photographs by Davida Johns; foreword by Raymond Arsenault and Gary R. Mormino.
p. cm. — (Florida history and culture series)
Includes bibliographical references (p.) and index.
ISBN 0-8130-2592-3 (c: alk. paper)
1. Agricultural laborers—Florida. 2. Migrant agricultural laborers—Florida. I. Title. II. Series.
HD1527.F6 R55   2003
331.7'63'09759—dc21   2002029908

The University Press of Florida is the scholarly publishing agency for the State University System of Florida, comprising Florida A&M University, Florida Atlantic University, Florida Gulf Coast University, Florida International University, Florida State University, University of Central Florida, University of Florida, University of North Florida, University of South Florida, and University of West Florida.

University Press of Florida
15 Northwest 15th Street
Gainesville, FL 32611–2079
http://www.upf.com

This book is dedicated to all the farmworkers, past, current and future, and their families.

Without them, the cornucopia Americans expect in their stores and on their tables would not be possible.

Contents

List of Illustrations	ix
Foreword	xiii
Preface	xv
Acknowledgments	xvii
Prologue	1
1. Moving with the Crops	20
2. Wages	47
3. Housing	72
4. Education	88
5. Health and Safety	105
6. Pesticides	126
7. Immigration	151
8. Family Life	178
Epilogue	189
Notes	193
Bibliography	199
Index	203

Illustrations

Prologue

1. Workers get on the bus in fog 2
2. Typical trailers housing farmworkers 3
3. A couple picks beans near Lake Okeechobee 6
4. Couple with laundry 9
5. Greg Schell and worker 12
6. Woman picks strawberries 14
7. Fern cutter under a tarp 15
8. Bean crates in a field 17
9. Woman hanging laundry 18

Chapter 1. Moving with the Crops

1.1. Ignacio Uribe and his wife, Antonia Tello 21
1.2. More members of the Uribe family 23
1.3. Sylvia Medina in a field 26
1.4. Workers combine cucumbers 27
1.5. Woman with bucket of tomatoes 29
1.6. Workers in packinghouse 30
1.7. Worker heaves cukes onto a truck 31
1.8. Workers leave on the bus before dawn 33

1.9. Crates of beans in a field 35
1.10. Workers with jackets leave the bus at day's end 37
1.11. Worker covers field with plastic 38
1.12. Orange grove worker on ladder 41
1.13. Woman picks kumquats 42
1.14. Catalano wraps ferns 43
1.15. Tractor in sweet potato field 45

Chapter 2. Wages

2.1. Workers in line with tomato buckets 49
2.2. Bean-field scale 50
2.3. Ramona Zarata sells tacos 53
2.4. Lucas Benetiz and Immokalee workers in office 54
2.5. Woman picks tomatoes in long rows 56
2.6. Lucas Benetiz and workers in front of Immokalee office 59
2.7. Workers cut cabbage and put it onto conveyor belt 61
2.8. Men pack zucchini 62
2.9. Woman picking strawberries 65
2.10. Workers board the bus to the fields 68
2.11. Citrus picker dumps oranges 70

Chapter 3. Housing

3.1. Old house on hill 73
3.2. Work clothes hung to dry in a fence 74
3.3. Worker in front of a trailer 76
3.4. Woman cooks tortillas 78
3.5. Trailers at field's edge 80
3.6. Apopka cabins 81
3.7. Kitchen with child and cooler 84
3.8. Woman with ringer washer 86

Chapter 4. Education

4.1. Women study at Beth-El Mission 90
4.2. Tutors at Farmworkers Self-Help 95

4.3. Margarita Romo teaching "Dream Team" 97
4.4. Padron children 98
4.5. Woman picks strawberries with her daughter 101

Chapter 5. Health and Safety

5.1. Pierson labor camp with garbage 106
5.2. Two field toilets on wheels 108
5.3. Three field toilets without wheels 108
5.4. Fungal infection of feet 110
5.5. Woman stoops to plant cucumbers in plastic 112
5.6. Woman in clinic 115
5.7. Trash outside trailer 117
5.8. Mother and daughter pick tomatoes 120
5.9. Man sharpens knife 123
5.10. Woman wears makeshift poncho 124

Chapter 6. Pesticides

6.1. Pesticide containers and tank 127
6.2. Children at field's edge 128
6.3. Boards inform workers 131
6.4. Workers work around methyl bromide 132
6.5. Woman wears a pesticide company hat 135
6.6. Woman washes squash in packinghouse 136
6.7. Tractor mows stubble 139
6.8. Alfredo Baheña and fern cutter 141
6.9. Fern cutter 142
6.10. Margie Lee Pitter 146

Chapter 7. Immigration

7.1. Rainbow House sign at Farmworkers Self-Help 152
7.2. Haitians pick beans 154
7.3. "Her Soil Is Her Fortune": Belle Glade sign 164
7.4. Belle Glade loading dock 165
7.5. Belle Glade tenements 167

7.6. Haitian couple in fields   168
7.7. Greg Schell and Belle Glade workers   173
7.8. Woman holds her children   174

Chapter 8. Family Life
8.1. Man and child share supper   181
8.2. Children study at Beth-El Mission   182
8.3. Father Ramiro Ros leads the congregation at Beth-El Mission   183
8.4. Farmworkers at Beth-El Mission receive holiday groceries   185
8.5. Woman receiving Christmas chicken at Beth-El Mission   186

Foreword

Florida's Farmworkers in the Twenty-first Century is the twenty-third volume in a series devoted to the study of Florida history and culture. During the past half century, the burgeoning population and increasing national and international visibility of Florida have sparked a great deal of popular interest in the state's past, present, and future. As the favorite destination of countless tourists and as the new home for millions of retirees and other migrants, modern Florida has become a demographic, political, and cultural bellwether. Unfortunately, the quantity and quality of the literature on Florida's distinctive heritage and character have not kept pace with the Sunshine State's enhanced status. In an effort to remedy this situation—to provide an accessible and attractive format for the publication of Florida-related books—the University Press of Florida has established the Florida History and Culture Series.

As coeditors of the series, we are committed to the creation of an eclectic but carefully crafted set of books that will provide the field of Florida studies with a new focus and that will encourage Florida researchers and writers to consider the broader implications and context of their work. The series will continue to include standard academic monographs, works of synthesis, memoirs, and anthologies. And while the series will feature books of historical interest, we encourage the submission of manuscripts on Florida's environment, politics, literature, and popular and material culture for inclusion in the series. We want each book to retain a distinct personality and voice, but at the same time we hope to foster a sense of community and collaboration among Florida scholars.

On Thanksgiving evening, 1960, CBS television aired "Harvest of Shame." Juxtaposing the swelling notes of Aaron Copeland's "Appalachian Spring" with the stern narration of Edward R. Murrow, the CBS documentary shocked millions of overfed Americans expecting holiday fare. "This scene is not taking place in the Congo. It has nothing to do with Cape Town or Johannesburg. No, this is Florida." So began Murrow, standing in Belle Glade, Florida. "This is the way humans hired to harvest the food for the best-fed people in the world get hired."

In the tradition of Edward R. Murrow's "Harvest of Shame," Alec Wilkinson's *Big Sugar*, and Cindy Hahamovitch's *The Fruits of Their Labor*, Nano Riley has advanced our understanding of the men, women, and children who till and harvest Florida's fields and groves. Riley takes readers into the lives of migrant labor: a world of backbreaking work, hazardous travel, and dangerous pesticides.

Florida Farmworkers portrays places far removed from the glitzy tourist brochures that glamorize the Sunshine State. To understand the lives of Margarita Romo and Lucas Benetiz, Riley examines the hardscrabble communities of Indiantown, Belle Glade, Dade City, Wimauma, Ruskin, Immokalee, and Gretna. If the conditions are bleak, the stories are stirring. If, in 1960, migrant workers held little power, Riley acknowledges that a cadre of dedicated activists now fights for worker rights: Florida Rural Legal Services, Farmworker Self-Help, Farmworker Association of Florida, Migrant Farmworker Justice Project, Farmworker Ministry, and Coalition of Immokalee Workers. *Florida Farmworkers* deserves to be read by every Floridian who wonders how Frostproof oranges, Belle Glade beans, and Ruskin tomatoes come to the table.

Raymond Arsenault and Gary R. Mormino
Series Editors

Preface

Florida is one of the major agricultural areas in the United States, though most people consider tourism the state's most important industry. Yet agricultural workers have traveled here since the Civil War to harvest the winter crops.

This book takes a closer look at the issues involving both the southeastern farmworkers who travel through Florida annually and those who make the state their permanent home. A comprehensive history of farmwork in Florida would fill several volumes. This book, however, intends to offer an overview; it is an attempt to make people aware of these invisible workers and their lives at the beginning of the twenty-first century. The photographs tell the story, and the many interviews with farmworkers and the advocates who work with them provide basic information about issues unique to Florida's agriculture and its farmworkers. It is a window into their hidden world.

Acknowledgments

This book began with an idea and a curiosity about the lives of these invisible farmworkers, and it took shape as a thesis. I would like to thank the many farmworkers and advocates who shared their time with me, describing their lives and concerns. I would also like to thank my professors at the University of South Florida, who carefully guided me into the history of this subject: Dr. Raymond Arsenault, who inspired me to look deeper into Florida history; Dr. David Carr, who encouraged me to study medieval agricultural history; and Dr. David McCally, for his information on the Everglades and the southern agricultural area of Florida and for reading the manuscript at its various stages. Finally, I would like to thank Peter Bramley, my partner of many years, for his patience and for walking the dogs while I finished this book. And many thanks to Susan Fernandez, formerly of the University Press of Florida, for her encouragement.

Nano Riley, Author

I acknowledge Margarita Romo of Farmworkers Self-Help in Dade City, who introduced me to the farmworker community throughout Central Florida.

Davida Johns, Photographer

Prologue

At the end of a rutted dirt road meandering across open fields in southern Hillsborough County stands a group of twenty or so ramshackle trailers. Scrubby brush and tall, waving Johnson grass borders the fields and obscures the little camp from the two-lane county highway. On the stoop of one faded yellow trailer, a skinny orange cat sits sunning itself next to a neat pot of red plastic flowers. Nearby someone has mended another trailer's broken window with masking tape, and instead of curtains a bedsheet stretches across behind the glass. Several other trailers in severe disrepair are padlocked, condemned, and ready for the junkyard. Here and there, a palmetto casts a bit of shade, and a few children's toys lie scattered in the dirt yards. Clotheslines stretched between the dwellings display jeans and other work clothes. At midday, there is no one around. The only people who come here are the farmworkers who live in the trailers and the grower who owns the surrounding fields and the trailers that he rents.

There are hundreds of these small enclaves across rural Florida, but few people ever see them. Tourists think of the Sunshine State in images of warm beaches, sultry breezes, glitzy hotels on palm-lined avenues, and Disney World, and they seldom venture off the interstate. Farmworkers, too, are seasonal visitors, but they come to pick Florida's winter crops of vegetables and citrus. Their communities remain invisible, even from the

1. On a misty spring morning in Immokalee, workers begin to congregate as early as 5 A.M. in a parking lot to await buses that transport them to the fields. They line up for particular buses driven by crew leaders. Many work for the same crew leader all week, but some may work for a different crew leader each day. At week's end, workers must collect their wages from each crew leader.

2. The sandy, rutted road leading to these trailers is miles away from any well-traveled road in Manatee County. Trailers like these provide typical housing for many migrating farm-workers. (Photo by Nano Riley)

two-lane highways that criss-cross the state. The fields where they work sit back from the roads, and the workers—miniature figures in the distance—are hardly noticeable behind the rangy hedges of elderberry and cottonwood trees. Interior Florida is a different world.

In 1960, Edward R. Murrow, the television journalist known for his trademark cigarette and his testy investigations, brought the farmworkers' problems to television in his landmark documentary "Harvest of Shame." When the program aired on Thanksgiving, it jolted the nation into an awareness of just where much of the food on America's tables originated. It was dramatic, and it showed the grinding poverty and instability of farmworkers' lives. Poignant interviews portrayed migrant workers as hopeless wanderers with a hard and futile future. Murrow's report began in Belle Glade, the small Florida town nestled in the reclaimed muck on the banks of Lake Okeechobee that is known as a major destination for eastern farmworkers. Across the country, Americans were shocked. It was the beginning of the 1960s, and a new social consciousness was taking hold.

In 1961, a citizens' committee in Belle Glade issued a response to Murrow's stark documentary in an effort to show that their community was not oblivious to the living and working conditions of Belle Glade's migrant population. The writers of the report described themselves as a "committee composed of local interested citizens of Belle Glade, Florida." Throughout the report, the committee attempted to rectify the negative impressions presented in the television program. The sixteen-page reply asserted that it was "submitted in the interest of fairness and to refute impressions created by the migrant report, 'Harvest of Shame,' presented over the Columbia Broadcasting System Television Network on November 25, 1960."[1]

The civic-minded residents of Belle Glade who wrote the report remain anonymous, but their mission was to take each injustice mentioned in Murrow's exposé point by point and offer their own observations on migrant living conditions in their community. The Belle Glade report addressed housing, education, health, welfare, and labor conditions of farmworkers in Belle Glade and surrounding Palm Beach County, and it discussed how growers were working with the government to correct any problems.[2]

"We are not 'Johnny come latelies' to this problem," declared the committee as it refuted Murrow's worst allegations. It denied the high cost of housing, touted the school system (both "colored schools" and "white"), discussed improvements in health care, and referred to the migrants as "free agents" who willingly come to Belle Glade because there is work. In the report's conclusion, the committee wrote that "there is one community that does not hang its head in relation to its contributions to migrants. An injustice has been done." Though the citizens' reply gave Belle Glade an improbably glossy image, it does offer interesting data on schools, health care, and other area institutions. It recognized that in 1955, between 15,000 and 20,000 migrants resided part of the year in Palm Beach County alone, more than in any other Florida county. It defined "migrant" as "an agricultural worker who crosses state lines one or more times during the year in search of work," while "seasonal workers" work the harvests near their homes. According to the report, 6,000 to 7,000 migrant workers spent time in the Belle Glade area, and of those, "over 80 percent are negro field workers."[3]

An earlier study commissioned by Governor LeRoy Collins in 1955, titled *Migrant Farm Labor in Florida: A Summary of Recent Studies*, remained unpublished until 1961. However, after "Harvest of Shame" put Florida under a microscope, state officials released the Collins study because of the "recent national publicity given the migrant problem." This broader report encompasses general concerns about farmworkers in Florida, from Seminole County in the central part of the state to the Everglades agricultural area in the south. This report, too, shows that most of Florida's field-workers in the 1950s were African American, though the presence of Mexican workers was increasing noticeably. This evolving demographic of farmworkers is characteristic of agricultural labor, which is usually composed of families displaced by calamitous weather, war, or disaster. Earlier, in the 1920s, these indigenous people of northern Mexico ventured over the border into the Rio Grande Valley, to become the Texas Mexicans. By 1953, these workers finally made their way into

3. Workers spend hours each day stooped in backbreaking positions in order to pick the vegetables that come to America's tables. Here a couple of people pick beans near Lake Okeechobee in the Everglades agricultural area, famous for its bean crop, which workers refer to as the "bean deal."

Florida's labor force, and most growers found them industrious and seemed pleased with their work, as the Collins study notes:[4] "They [the Texas Mexicans] work in both the vegetable areas and in citrus, and are generally regarded in both as superior to the usual run of domestic Negro migrants in that they maintain a higher living standard, they take better care of the camp property and cause less trouble with requests for salary advancements."[5]

The migrant labor study also mentions the presence of Puerto Ricans, Bahamians, and Jamaicans, but they composed only a third of the black population. White workers were the smallest group of farmworkers in Palm Beach County in 1955. The workforce was transforming once again, finding new laborers willing to do the work most native-born Americans refused.[6]

Farm Labor in America

America founded its economy on the backs of agricultural laborers. When the first Europeans arrived at the Virginia colony of Jamestown in 1607, tobacco was the cash crop, produced with indentured European servants and a few African slaves. At the beginning of the twenty-first century, agriculture in the United States still relies heavily on migrant and seasonal farmworkers for labor-intensive crops, especially in Florida, where produce thrives year-round. This human labor force has been indispensable to the success of American agriculture.

After the Civil War, "truck farms" emerged in southern New Jersey as suppliers of vegetables, fruits, and berries for the growing urban centers of New York, Philadelphia, and Newark. East Coast farmers could truck their produce to market in wagons, but they needed extra laborers to help them during harvesttime, laborers who would leave when the work was done. In the early days, the workers were usually African Americans from the South, but as the influx of immigrants grew toward the end of the nineteenth century, Italians, Chinese, and Filipinos joined the agricultural workforce. As technology improved, allowing produce to be shipped long distances in refrigerated train cars, the truck farms spread down the East Coast, bringing with them the inevitable supply of migrant farmworkers.[7]

In the early 1930s, during the Great Depression, many refugees from the drought-stricken Dust Bowl entered the migrant stream, toiling alongside other farmworkers. While Georgia and the western states reeled from depression economics, the boll weevil, and the Dust Bowl, Florida's farmers grew wealthy as their vegetables flourished in the rich muck along the exposed banks of Lake Okeechobee. Because Florida's farmers were so successful at raising several crops a year in the temperate climate, they needed more hired workers for longer periods. There was lots of work, and hundreds of out-of-work people flooded into Florida for the winter seasons, seeking any job that would feed their families. Labor became the growers' greatest expense. Though growers hired hundreds of workers, they paid less than did farmers in states with shorter growing seasons, who needed labor for only a few months each year.[8]

With the beginning of World War II, many white workers moved out of farmwork into higher-paying jobs in defense factories, but this option was open to few African Americans because of segregation. In order to maintain sufficient farm production during the war, when so many in the workforce were called up for military duty or found better jobs related to the war effort, the government began the *bracero* program, which allowed farmers to import workers from Mexico for specific time periods. The *bracero* program, also called the Mexican labor program, began in 1941 and lasted until 1963. During the twenty-two years this program was in place, over 4.5 million Mexican nationals came, under contract, to work on American farms. At the same time, workers from the Caribbean were imported to fill positions in Florida. These island laborers—from Jamaica, the Bahamas, and Puerto Rico—were called "offshore workers" to differentiate them from the Mexican workers.[9]

At present, three major north-south migrant streams exist in the continental United States. Migrants based in southern California journey to northern California, Oregon, and Washington. Those based in Texas and Arizona usually travel up the Mississippi Valley to Ohio, Michigan, Indiana, and Illinois. Still others follow the East Coast, from southern Flor-

4. This couple from the islands carries laundry home in the traditional way. Belle Glade is home to many workers from the Caribbean, who came to Florida to pick the fields and decided to make America their home.

ida through Georgia, the Carolinas, Maryland, Delaware, and New Jersey into New York and New England.[10]

Florida's Farmworkers

Florida's first hired field-workers, in the mid-nineteenth century, were Native American descendants of Creeks and Cherokees who had fled to South Florida to avoid the Indian removal of the 1830s. By the end of the nineteenth century, the Bahamians, who regularly fished and sponged in the Florida Keys, outnumbered the Native Americans. Soon these island people, plus the African Americans from the Cotton South, formed the face of migrant labor in Florida. Georgia's boll weevil infestation doubled the hardships of southern farmers in the late 1920s, creating a mass migration of sharecroppers and tenant farmers who could no longer eke out a living in their cotton fields. Thousands of Georgians, both black and white, joined other out-of-work laborers to follow the harvest from Maine to Florida. But while the African Americans toiled in the fields, the white workers usually had jobs in the packinghouses. If white laborers did pick alongside blacks in the fields, their pay was sometimes twice that of the African Americans. Around Belle Glade, green beans were the most popular crop because they matured quickly, offering the planter greater profits. Workers referred to the crop as the "bean deal."[11]

Beginning in the 1950s, Hispanics from Mexico and Puerto Rico joined the African Americans to dominate the ranks of Florida farmworkers. Jamaican, Haitian, and Dominican men came in as offshore workers to harvest the fields of sugarcane growing thick around Lake Okeechobee. Today, most foreign workers flood into Florida from Mexico and Central America, but some still come from the Caribbean islands of Haiti, Jamaica, and the Dominican Republic.[12]

Now the newest players in Florida's farm labor force are political refugees from Central America and southern Mexico. Beginning in the mid-1980s, Salvadorans and Guatemalans made their way north, fleeing the death squads that ravaged their countries during the 1970s and 1980s. These immigrants from Central America and southern Mexico, from

Chiapas and Oxaca, are not the Mexicans who came in the 1950s and 1960s from northern Mexico's high plains. They are of Mayan descent and often speak several varied dialects but little or no Spanish. The majority of these newest arrivals are young men who come alone to find a better life in the United States. Here they have a chance to make money and to realize the hope that they may move up the ladder from farmwork to a steady service job.

Many of these single men are in the United States illegally, having either sneaked across the border on their own or arrived with a "coyote," a man who smuggles those without proper papers into the country. These men work hard, stay out of sight of authorities, and often pay several hundred dollars for the phony documentation they need to work. Some farmers do not even question the legal status of workers, especially those who arrive with a crew boss. There are no exact figures on the number of these lone workers, but this group is different from the Mexican workforce of earlier decades.

The Mexican workers who have been coming since the 1950s retain close family ties with relatives in their hometowns in northern Mexico, usually returning there once a year to visit. If single, these workers often send money home to relatives in Mexico and hope to bring their families to the United States with them once they are established and have a green card. It is these families who have "settled out" in rural areas of Florida, traveling little or not at all, and now find themselves working steadily in one area for local farmers.[13]

Besides the legal and illegal workers who come to the United States to work in the fields, the government's H-2A program allows labor contractors to provide a grower with temporary foreign workers just for the harvest period. When the work is done, these "H-2" workers, as they are called, return to their own countries. Before 1996, the sugarcane farmers used hundreds of H-2A workers from Haiti and Jamaica annually to cut the sugarcane with machetes. But since the mid-nineties, improved technology and the diminishing soil surrounding Lake Okeechobee made mechanical harvesting machines less expensive than humans. With the hard limestone of the Everglades region covered by only an inch or two of

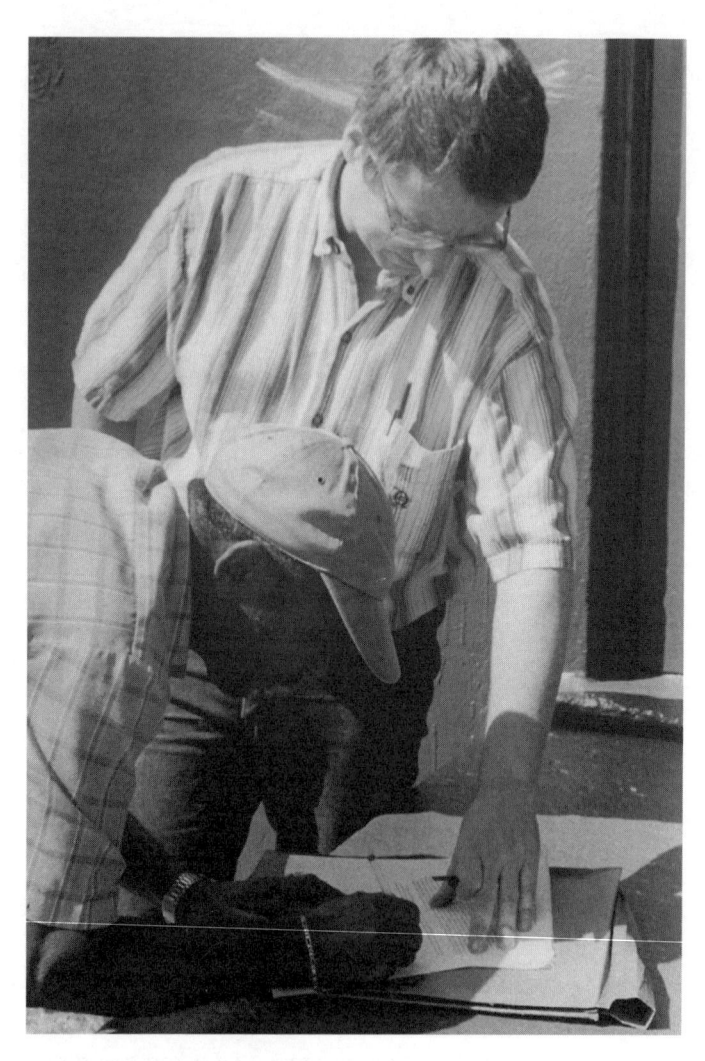

5. Greg Schell, a lawyer with the Migrant Farmworker Justice Project in South Florida, shows a Haitian worker where to sign to receive a settlement from a suit against a grower for withholding wages.

soil rather than the thick, soft muck of earlier years, "Big Sugar" can now get machines in to cut the cane.[14]

Problems of Farmworkers

Migrant and seasonal farmworkers are one of the most underserved and understudied occupational populations in the United States, even though they are working in one of the most hazardous occupations in this country. In Florida, farmworker population estimates range from 200,000 to 350,000, depending on whether or not one includes undocumented workers. In 1998, the agriculture industry accounted for 780 deaths and 140,000 disabling injuries nationwide. Agriculture workers had the second-highest death rate among the major industries.[15]

Edward R. Murrow's powerful documentary struck a chord in many Americans, inspiring reformers in the 1960s to take genuine interest in the problems of farmworkers. In 1963, Congress passed the Farm Labor Contractor Registration Act (FLCRA), requiring all labor contractors to register with the government, provide accurate wage records, and disclose working conditions. In 1962, a young Chicano named Cesar Chavez founded the National Farmworkers Association, which later became the United Farmworkers, a powerful AFL-CIO affiliate that guards farmworkers' rights. During the 1960s, many people also assumed technology would soon reduce the dependence on manual labor, but tree-shakers for dropping fruit and bean-harvesting machines have not emerged as viable alternatives to a human workforce.[16]

But in spite of Chavez's success in California, in Florida unions are little help. Since Florida is a right-to-work state, which means that anyone, whether a union member or not, can work, unions wield little power to increase employees' wages. Earning only the barest minimum wage, farmworkers are often forced to live in squalid, crowded housing with poor sanitation. Such close conditions encourage the quick spread of disease, and tuberculosis still is a problem in some communities. These laborers also perform hazardous, backbreaking labor, stooping in

6. Strawberries are one of Florida's major crops. This woman picks the prickly berries near Homestead. After filling her boxes, she will carry them yards away to the field's edge, where the crew boss tallies her harvest.

7. Fern cutting is a dream job for most farmworkers, because the shade-loving ferns thrive under tarps, which make the work cooler. Still, there are hazards in the form of pesticides and the occasional snake that slithers through the fern patch.

the fields for hours as they harvest vegetables that have been sprayed with dangerous pesticides.[17]

In 1998, the National Institute for Occupational Safety and Health (NIOSH) ranked "farmworker supervisor" as the fourth most deadly occupation in the United States. Other problems result from the work's mobility, such as farmworker children changing schools each time the family moves, making it difficult to learn. Because of the ever-present poverty, there may be problems at home, including domestic violence.[18]

Because of their mobile lifestyle, there are no exact statistics for the farmworker population in the United States. Many move often, and others avoid authorities, especially if they are *sindocumentos* (without documents). Confusing the count is the problem that there is no differentiation made between migrant and seasonal farmworkers in most agricultural surveys. In 1986, when the government enacted the Immigration Reform and Control Act, the U.S. agricultural workforce was estimated to be about 6.5 million, 5.4 million of whom lived on farms and 1.1 million of whom were hired workers. Estimates indicate that as many as 5 million migrant and seasonal agricultural workers live and work in the United States annually, but most statistics generally underestimate agriculture's dependence on hired workers.[19]

Migrants may remain uncounted for several reasons. There is no uniform definition of migrant and seasonal farmworkers among government agencies. Language barriers can make information they give inaccurate; the seasonal nature of the work requires them to move often; and the large distances between camps or farms in rural, often remote, areas make the workers inaccessible. Currently the only national reporting system that tracks farmworker health data is the Migrant Student Record Transfer System maintained by the Office of Migrant Education of the U.S. Department of Education. This computerized system contains the health and academic records of children of migrant farmworkers in the United States and Puerto Rico, but there is no such collection of national health data on adult farmworkers.[20]

It is also true that farmworkers receive benefits from many government and private social service organizations. Because they do not have

8. Workers sometimes sit on milk crates to pick beans in South Florida. Here the stacked crates bulge with beans, as pickers await the truck to take them to the packinghouse, where the beans are washed and prepared for market.

9. This woman raised a family while moving with the crops. This is just one place she stayed in Central Florida. Here she hangs the family's laundry outside this dilapidated house that sits on a grower's land at the edge of the fields.

steady employment and often live below the poverty line, they may qualify for free clinic services, dental care, and day care, as well as food stamps and rent subsidies. But sometimes these programs backfire, and a year of good earnings can reduce or eliminate these benefits for some. Also, most of the current programs are patterned on earlier models developed when families traveled together, but they are not geared to the men traveling alone who form the majority of today's farm labor force. For those whose documentation is in doubt, these programs do not exist.

Though farmworkers have their share of problems, they also spend many happy times in the company of their extended families. Often three generations live and travel together, providing continuity in an otherwise constantly changing world. They celebrate birthdays and other family occasions, and the Mexican workers often have community festivals for special holidays, such as Cinco de Mayo, a celebration gaining popularity in the United States. Churches and other advocacy groups also sponsor events to raise funds for the workers who may need help meeting the basic needs of their families. In late summer, when work is slow, many laborers return home, to Mexico or the Texas border towns where they were born, to visit relatives and enjoy a vacation.

Farmworkers are often portrayed as a downtrodden underclass. They are a frequent subject of journalists, whose stories may shock comfortable Americans accustomed to a different standard of living. But the plight of these workers will remain as long as most American citizens refuse to do this backbreaking labor. Hard work is a fact of the lives of farmworkers, and most of these laborers look upon themselves as fortunate to have work and to be in the United States. Though much has been accomplished by hard-hitting reports, such as Edward R. Murrow's, it remains that people must recognize the needs of these laborers and the hazards they endure in order to put food on our tables.

CHAPTER 1

Moving with the Crops

Traveling

As they follow the crops working harvest to harvest, migrant farmworkers and their families face many challenges. Farmers only hire extra help for brief periods when planting or harvesting crops, so steady, long-term work is seldom a possibility. In addition to an uncertain social and economic future, these workers must also contend with unreliable transportation. Since most farmworkers cannot afford dependable vehicles, and gas to travel is expensive, many workers rely on crew bosses with trucks or buses to transport them to and from the fields and even from state to state. Traffic accidents involving farmworkers are common, especially on rural roads. The thick, predawn fog that blankets many Florida country roads can cut visibility to a minimum. Add to that an old truck with poor brakes, and the combination can be deadly.

For the school-age children of these traveling families, it may be difficult to get a good education or to form lasting friendships outside of their own community. Each place they land is temporary, a home for only three or four weeks. The dwellings where they stay are, at best, minimal. Things most American families take for granted, such as televisions,

1.1. Ignacio Uribe and his wife, Antonia Tello, are from El Parajito, a small Mexican mountain village. They have migrated for over thirty years to Texas, Florida, the Carolinas, and Ohio. Now in their mid-sixties, they had fourteen children. Two died at birth, one girl died at five years, and a son died at two years. They also have two adopted children and twenty grandchildren. Ignacio began migrating illegally and alone, earning money for months at a time in the United States while his family tended the crops in Mexico. Later, he would bring two or three teen-aged sons. Since he did not attend school, the youngest son began migrating at sixteen to work in the orange fields near Sebring. Now more of the family lives in the United States than in Mexico. One son lives permanently in South Carolina, where he arranges housing and transportation to the fields for workers who continue to migrate. Most of the Uribe family travels back to the little village of about 300 or 400 residents for the Christmas holidays. The drive from El Parajito to Central Florida takes two to three days.

computers, video games, and even working refrigerators and ovens, may not be part of farmworkers' wandering lives.

In *Children of Crisis,* a landmark study of rural laborers of the 1960s, psychologist Robert Coles noted that many of the migrant children he met lived by the rhythms of nature. They often told time by the sun and the moon rather than in minutes and hours. Coles also found that children of farmworkers often assumed many responsibilities around the home at very early ages. They frequently cared for each other while both parents were at work, and they learned to make food when they were as young as two or three. These children were also adept at talking to outsiders and authority figures, who could cause trouble for their families. However, Coles discovered that most migrant workers, especially the families, yearn for the permanent work and permanent homes that would give their lives a sense of stability.[1]

Today, even as farming becomes increasingly mechanized, the human touch is still a necessity for picking most of the delicate fruits and vegetables that come to our tables. Workers who pick tomatoes in southern Florida in the fall and spring travel north up the eastern migrant stream in summer to pick peaches in Georgia. Later in the season as the northern crops ripen, they head to New Jersey and Pennsylvania, while some travel as far as West Virginia and the midwestern farm states of Ohio and Illinois to pick apples and other fruits and vegetables. In September, when the northern harvest is done and Florida's farmers are ready to set the first fall crops, the workers turn south, drifting back to begin the cycle again. During the winter months when snow covers the northern fields, Florida's winter crops keep workers busy.

Several families may often travel together, working and living in groups. Sometimes families return to the same farms year after year to pick the same crops. Constant moving is a way of life for these people, but because several generations often stay together, there is an element of continuity in their nomadic lives. Though families played a large role in the farmwork of past years, many of them have "settled out," meaning they have stopped traveling and now work locally near permanent homes. With the help of the social and educational programs introduced

1.2. Whole towns sometimes migrate together from Mexico, traveling together for safety and sociability. Besides often living and working together, they also watch out for one another. These extended families of all ages add a degree of continuity to a life that constantly changes. These are more members of the Uribe family from El Parajito, Mexico, photographed outside a packinghouse near Dade City.

since the 1960s, many of these workers and their children have moved away from backbreaking field labor into more skilled jobs in farm work or into other work altogether.[2]

Today the workforce is in transition as the trend in farming shifts toward corporate farms funded or owned by agricultural giants. These large farms prefer to recruit men traveling without families to do fieldwork because it is more cost-effective. Mexico provides most of these laborers, who stream across the borders, legally or illegally, in search of work. Crew bosses or growers can house them in dormitories or crowded trailers, and there is no need to provide for nonworking family members, such as children or the elderly, as farmers did twenty years ago. Sometimes the men traveling alone are fathers who leave their families in Mexico until they can establish a definite work pattern that enables them to bring their wives and children into the country. Other farmworkers are young, single men who hope to find that pot of gold in America. These men are the ones who may suffer abuse from crew bosses, who often cheat them out of their wages. Many workers who entered the country illegally paid a coyote, or smuggler. These *sindocumentos,* as workers without proper documentation are called, speak little English and depend on their crew bosses for everything. They usually avoid authorities and seldom complain about wages or bad treatment because they fear deportation.[3]

In 1977, 75 percent of Florida's migrant and seasonal farmworkers were single men, a statistic that shows the beginnings of today's trend in agricultural labor. Even though the sugarcane industry now uses mechanical harvesting machines and no longer imports offshore cane cutters from the Caribbean, the majority of migrant workers in Florida today are men traveling alone.[4]

The changing demographic at the beginning of the twenty-first century shows that more families are settling out, particularly in Florida, where the climate allows year-round work in agriculture. The children of these families can often remain in the same school, which creates more stability in their lives. The adults benefit from social and educational programs available to them when they remain in one place, and they enjoy the income from steady employment.

Sylvia Medina

Sylvia Medina is a crew boss these days, but she remembers when she was moving around the country as a migrant. She traveled down to Homestead for the winter vegetables, up to South Carolina and Virginia for the summer harvest, and finally back to Wimauma, in southern Hillsborough County. The Medinas settled out in 1990, and she and her husband of twenty-six years, Narcisso, own a large trailer on a sizeable lot along the rural road that runs between the tiny towns of Balm and Wimauma. The double-wide has a satellite dish perched on top that allows the Medinas to tune in to stations from the Mexican border; sounds of a Spanish program drift from a back-room television someone is watching. They also own several trucks, along with a house in Florida City near her father's farm in Homestead. Sylvia and Narcisso Medina have three daughters, three sons, and nine grandchildren, and the close-knit family often works together in the fields near their Wimauma home. Farmwork is the only work that any of them knows.

At forty-four, Sylvia Medina is happy that migrant work is in her past, and she talks about how difficult it was on the road. Her English is good, and she likes to talk. Her husband is more stoic and speaks mostly Spanish. The Medinas' crew works mostly for Saffold Farms in southern Hillsborough County.

"I started working in the fields when I was about fifteen. I worked with my family. I was born in Brownsville, Texas, so I am an American citizen.

"It was always hard work. Many of [the bosses] watch over you. They're very strict. One boss used to time us. He'd tell us there were so many beets in this row—do it in an hour; and we had to do that. I did tobacco and sweet potatoes, and that's a hard job. The dirt gets up in your nails—it was the hardest job I've ever done. I picked tomatoes in South

1.3. Sylvia Medina has been a crew boss with her husband, Narcisso, for fifteen years. Many workers in her crew are family members. The Medinas are popular bosses who work alongside their crew. They own several trucks and contract with farmers to take produce to market. Here Sylvia supervises workers picking cucumbers. Workers pick fields several times as crops ripen. This is the last time they will pick this cucumber field.

1.4. Most workers start work in the early morning before the sun is up. Here Sylvia Medina's crew gathers the last cucumbers from this harvested field. The mist still hangs along the edges of the fields when the workers begin, just as the sun rises.

Carolina. You have to wear gloves and long sleeves; otherwise your arms turn green where the stems rub you. I also picked lots of okra, and it is real prickly."

In 1985, the Medinas became crew bosses, but they continued to travel the migrant stream for several years, hauling their main workers with them.

"When we traveled, we must take everything. We had eight trucks—produce trucks, and even the water truck. We took them all—the trucks and the crew," she says.

Medina says that her crew can blossom to as many as ninety pickers as the season peaks for tomatoes, cucumbers, cabbage, and peppers in Central Florida.

"They all have their green cards or they were born here [in the United States] near Brownsville," she maintains. "My sister-in-law works with me, and many on my crew are cousins. Most of the time I have the same crew of about thirty workers, except at the height of the season when we hire extras. When we're not picking, my crew can work for anybody. Sometimes they work in nurseries," she explains. "Most of the workers rent from the grower.

"This year, just inland from Wimauma the fields froze, but we were lucky—no freeze," says Medina. "When it freezes, it puts people out of work, and then you have to go where the crops didn't freeze."

In the fields, Medina is a skilled professional. She walks quickly among the rows of cucumbers, picking only the ones best for the grocery store. She and her crew carry red plastic half-bushel containers. Each worker picks until the container is full of cucumbers, then hands the container up to a worker who dumps it into the truck driving slowly along a raised road that runs through the fields.

"Cukes can't be used if they're too yellow or too small," she explains. "We pick each field three times. We also plant the fields. You can plant peppers in little flats, but cucumbers are hard to plant. You must put the cucumber seeds in the holes by hand, two seeds to a hole.

"I keep track of everyone's hours. I put everybody's name down and how long they work," she says. "We'll be here 'til about three this after-

1.5. This woman carries her bucket of tomatoes toward the truck that drives slowly through the fields alongside the pickers. Long sleeves and gloves protect her arms and hands from irritating red tomato stains, and the hat shields her from the sun.

1.6. After a long day spent harvesting in the sun, workers often continue until midnight preparing vegetables in the packinghouses and then rise at dawn to return to the fields. Here workers grade peppers for packing near Wimauma, in Hillsborough County.

1.7. A worker delivers a bucket of cucumbers to the field truck. First, the workers stoop in the fields to harvest the vegetables; then they heave the bucket up to the truck, where someone dumps its contents into the bins. The man who takes the bucket will tally the number of buckets each picker delivers. When the truck is full, it takes the produce to the packinghouse, and another takes its place in the field.

noon. We started at eight-thirty—we got a late start this morning because of the fog."

When all the fields are picked for the day, the crew heads back to the packinghouse near the Saffold Farms offices. The packing shed, a large, barnlike structure, is outfitted with tables for sorting and packing the vegetables for market. Overhead lights hang down on long cords so workers can see clearly as they clean and sort the produce. After a long day in the fields, farmworkers often work late into the night inspecting the produce and packing it in boxes to be ready for the trucks in the morning. Outside, where several scrawny dogs lounge in the dirt, a large truck bears the Medina name. The Medinas own it and make extra money taking produce to the local wholesale produce market.

The Medinas have been fortunate. They no longer have to move around, though they go to their other house near Homestead every few months, often just to help Sylvia's father, who grows vegetables nearby.

"My father has 100 acres in Homestead, where he grows okra, zucchini, yellow squash and crowder peas. He pays the pickers $2.50 per box, and he gets about $30.00 for it. I used to pick fifty boxes by 11 A.M. He doesn't have his own packinghouse, so we have to use someone else's to pack up the vegetables."[5]

Transportation

Most migrant farmworkers do not own reliable trucks or cars, yet they still travel for miles to meet the next harvest in whatever they can afford to drive or in the back of a crew boss's van. Traveling in these unsafe, often overcrowded pickup trucks makes farmworkers vulnerable in even minor mishaps. Statistics from the Bureau of Labor for the southeastern United States list 199 farmworkers killed in work-related incidents in 1999. Of those workers, 22 percent died in highway accidents. In 1998, the National Institute for Occupational Safety and Health (NIOSH) issued a report on occupational safety that found traffic-related accidents are the most common cause of on-the-job fatalities, leading to more than 1,000 deaths per year.[6]

1.8. An old school bus takes these workers to the fields in the pre-dawn hours.

There have been many highly publicized incidents of fatal highway accidents involving farmworkers, and these are usually due to vehicle problems. One such wreck in Central Florida occurred because of a van's disrepair. In a case that tested the rights of farmworkers in 1995, David Moody, a former Florida farmworker, testified for agricultural workers on Capitol Hill before the House Economic Worker Protection Committee. Gregory Schell, a lawyer with Migrant Farmworker Justice Project, a nonprofit agency in Belle Glade, and a longtime advocate for farmworkers, also spoke before the committee as Moody's lawyer.

Moody testified that after losing his job with Eastern Airlines as a mechanic during their 1990 bankruptcy, he was living in Camillus House, a homeless shelter in Miami. When Willie Lee Simmons offered him work in a warehouse near Orlando, Moody eagerly accepted. Simmons also picked up several other homeless men from the shelter and piled them into his brown Chevy van. The van's rear seats had all been replaced by boards stretched across cinder blocks.

The warehouse turned out to be a nursery. Because Moody was a mechanic, Simmons hired him to drive the men fifty miles a day from a labor camp in Pomona Park to the nursery. Simmons's van was in a dreadful state of disrepair, according to Moody. Though he did not intend to stay, Moody testified that he found himself trapped as Simmons kept part of his weekly pay for meals, cigarettes, and beer sold on credit. One early morning in a heavy fog, Moody had an accident, killing four of his fellow workers. He then discovered that Simmons had no insurance on any of his vehicles and had never bought the worker "B" compensation insurance required by law. There were no funds for medical care available for any of the victims.

Moody filed suit against White Rose Farms and Simmons under the Migrant and Seasonal Agricultural Worker Protection Act of 1986 and received a small settlement from White Rose. It seems that Simmons had several citations for unsafe vehicles and violating laws governing farmworkers' rights. Simmons spent eleven months in a county jail for his violations; shortly after his release, he was back at work with his son.[7]

In 1977, about 57 percent of farmworkers had working cars, while the

1.9. Beans are packed as they are picked, and the crates are weighed right in the field. The bus that brought the pickers to this field sits nearby on a dirt road, a passage running through several acres of beans. The scale is there so workers can be sure that each crate weighs the same.

remaining workers depended on crew leaders, friends, and relatives. Sometimes the rides were free, but often crew leaders charged the workers for gas. Today, families may own cars or trucks, particularly if they live and work in a regular area or only migrate seasonally. It is still rare for the single men making up the majority of migrant workers to own vehicles, which makes most of them completely dependent on crew bosses to get to the job.[8]

In 1991, seven workers near Lake Okeechobee died when a station wagon carrying them to work ran off a road in a sugarcane field and flipped upside down into a drainage canal. According to a highway patrol officer, the trapped men left claw marks on the inside of the vehicle and on each other as they tried to escape from the submerged car. In a news report of that accident, a sheriff's deputy commented that migrant workers are routinely transported in overloaded vehicles driven by unlicensed drivers.[9]

Even the families of farmworkers may be affected. In 1996 in Okeechobee, nine-year-old Epifania Campos-Cruz dashed across the highway in front of her house at 5:40 A.M. to tell the driver of a bus taking workers to the fields that her father was driving to work that day. As she crossed the road again heading home, she was struck and killed by a pickup truck. It was in December, so the morning was dark, the time when many traffic accidents involving farmworkers occur.[10]

The 1961 Belle Glade citizens' report written in response to the CBS documentary "Harvest of Shame" states that when traveling from state to state in the early 1960s, farmworkers used "the cheapest form of transportation available, . . . crew leaders maintain for this purpose a fleet of trucks which are usually old and broken-down." In the 1940s and 1950s, crew bosses often transported workers in closed-in trucks with doors chained from the outside, making it impossible for the workers to escape in case of fire or accident. Though the lack of safety was apparent, it was not until the 84th Congress authorized the Interstate Commerce Commission to regulate safety and comfort requirements for migrants transported farther than seventy-five miles across state lines that things changed. Still, the attitude expressed in the citizens' report was that too

1.10. After a day in the fields, these workers disembark from the bus in a parking lot in downtown Immokalee. The late afternoon sun is hot, but they carry their jackets because temperatures in the morning may be only forty degrees.

1.11. Farmworkers perform many different jobs. After tractors spray chemical pesticides and fertilizers on the newly prepared fields, another tractor attachment forms the soil into long, raised, square-edged mounds ready for planting. Another machine spreads plastic over the dirt, but a human is still required for the final work of securing the plastic with a covering of dirt along the edge, as this worker is doing. An old T-shirt placed under his cap provides some welcome relief from the hot Florida sun.

much regulation would deprive workers of a fair chance at employment.[11]

Traveling from home to the fields can be dangerous, and traveling the highways as farmworkers follow the crops is also risky in unreliable vehicles. But perhaps the most dangerous traveling farmworkers do is coming into the United States illegally, especially when smuggled in by a coyote. In August 1998, seven people believed to be illegal immigrants—five men, a woman, and a teenage boy—died in the California desert just north of the Mexican border, near Salton City. A spokesman for the INS said they found the bodies lying under a tree where they apparently had died of heat exhaustion. The temperature in the desert was 105 degrees in August, according to the National Weather Service. The INS official said that the smuggler had dropped the people off with no water and had ordered them to wait. According to the INS, people take increasingly difficult routes since the United States tripled its border agents at the most common crossings in "Operation Gatekeeper" in 1994. Since the attacks on the World Trade Center in 2001, the borders are even more closely guarded. Though this incident happened in the Far West, many Florida workers have entered the country in much the same way.[12]

The Crops

Farmworkers in the eastern migrant stream follow the warm weather when they travel. In the summer and fall they head north, and in the winter they come to temperate Florida. They seldom contend with ice and snow. They follow the crops as the harvests come near, either traveling in their own vehicles or riding with a crew boss. Some workers follow only certain crops, and some will work wherever there is work to do. They not only pick the crops, but they also prepare the fields for planting and maintain them before the harvest. Families who have settled out and no longer travel usually find work with one farmer year-round, plowing, mowing, and irrigating the fields. During harvesttime, they go from the fields to the packing sheds, where they often work until midnight or later preparing the vegetables for shipment to market.

One of the most popular vegetables grown in America is the tomato, and Florida produced 40 percent of the national crop in 1997, bringing in $487 million in revenue, second only to the profits from oranges. But the popular red fruit requires a lot of work. First workers shovel the fields and lay plastic as mulch to protect against weeds; then they plant the seedlings and tie them to the stakes. Tying tomato vines pays by the foot, and workers tie them each time they grow taller until they are ripe. After the harvest, the workers pull the stakes and the plastic from the fields so they can be prepared for the next planting. About a week before the workers return, the growers spray the highly toxic herbicide paraquat on the fields to kill the weeds. Signs posted on bulletin boards notify workers of the date the field was sprayed and the date they can safely reenter.[13]

Only citrus is more lucrative than tomatoes in Florida, bringing in $1,115,429,000 in revenue in 1997. Picking oranges is difficult work, requiring workers to climb tall ladders and balance bags weighing up to ninety pounds. Crew leaders usually provide the ladders for reaching the tops of citrus trees. Pickers prop them against limbs and climb to the treetops carrying a bag strapped around their necks and shoulders. Falls from the ladders are not uncommon, and pickers' hands turn orange from the skin and the sticky juice. Some citrus trees, especially grapefruit and lemon trees, have long thorns that can prick workers or catch their clothing.[14]

Because of Florida's warm winters, specialty crops thrive here as they do in Hawaii and California. Tropical fruits such as avocados and mangoes flourish in South Florida, especially along the east coast. Avocados are the largest specialty crop, providing nearly one million bushels of the buttery fruit annually, competing closely with California and commanding nearly $18 per bushel for an annual revenue of over $16 million in 1999. Mangoes also brought in about $1.5 million for about one million bushels in 1997, the latest year for which statistics are available. Picking avocados and mangoes offers dangers similar to those of picking citrus, requiring workers to climb ladders and fill heavy sacks.[15]

One interesting fruit that has gained popularity recently is the berry of the saw palmetto, which is used as a tonic for prostate troubles. Some of

1.12. Only oranges earn more for Florida growers than tomatoes, bringing in $1,115,429,000 in revenue in 1997. Picking oranges is difficult work, requiring workers to climb tall ladders and to balance bags weighing up to ninety pounds. Falls are not uncommon. Workers also have to contend with the thorns on the citrus trees, as well as the sticky fruit whose skin and juice turn pickers' hands orange.

1.13. Kumquats are one of Florida's tropical specialty fruits. Picking kumquats is tedious work. One hand selects and holds a branch loaded with kumquats, while the other clips the stem and drops it into a deep bag.

1.14. Catalano, nearly sixty, wears duct tape on his fingers to protect them as he cuts leather-leaf fern on a fern farm near Pierson, in Volusia County. The rubber bands on his wrists enable him to quickly bundle the fronds in an automatic action. Catalano is from southern Mexico, but he has lived in the United States since 1979. He no longer cuts 400 bunches a day, but he is still very fast and professional. He says he enjoys the work he has done so long.

the seasonal laborers in South Florida who do not want to travel pick saw palmetto berries in the summer. Gathering saw palmetto berries is tedious because harvesters usually work in undeveloped areas and must contend with the spiny stems of the palmetto fronds. But because the trees are not owned by anyone, pickers can sell them at market price and keep the total profit.[16]

And here in Florida it is not just produce that provides jobs for willing workers. The horticultural market has mushroomed in the last twenty years, offering work in the wholesale nurseries that flourish in the center of the state near Lake Apopka. Many workers here have settled out of the migrant stream and work year-round in the nurseries and greenhouses. Still, there is a specific growing season for certain crops, such as ferns, when temporary workers add to the permanent workforce. In Volusia County, the town of Pierson calls itself the "Fern Capital of the World." Here, growers devote acres of shady woodlands to different varieties of ferns, which are shipped to florists all over the world. In 1998, cut greens brought in a profit of over $98 million.

Though working as a fern cutter appears to be far more pleasant than stooped labor in the dusty fields or climbing ladders in an orchard, it still has its problems. Fern workers may suffer repetitive stress injuries from constantly bunching the ferns. As they cut the ferns, they gather the fronds in bunches of twenty-five and wrap them with rubber bands that are worn on their wrists like bracelets. A fast worker may bundle nearly 400 bunches of ferns a day, but it is still piecework, and it still pays only 21 cents per bunch, the same amount paid workers twenty years ago.[17]

Today, the social programs geared to families and legal immigrants offer an improved prospect for the education, health, and safety of the workers who toil in the fields. Yet workers and advocates can still cite many abuses and rights violations that exist in the marginalized world of farmworkers, especially for those on the bottom rungs, the illegals who make up the new majority of this workforce. In spite of the programs, farmworkers continue to travel from harvest to harvest in unreliable vehicles. Perhaps the major problem remains the low wages, for without

1.15. As a tractor uproots sweet potatoes in a field near Homestead, workers collect them from the dirt and pack them in bins.

enough income, these workers stand little chance of improving their situation.

Florida Crops

Though tomatoes are the most lucrative vegetable crop in Florida, other mainstays remain popular. In the northwestern panhandle counties near the Alabama border—Holmes, Jackson, Washington, and Gadsden—grow hearty crops of butter beans, field peas, watermelons, pole beans, squash, sweet corn, and tomatoes. The Suwanee River Valley and the northern areas of Starke, Brooker, and Lake Butler produce lima beans, snap beans, blueberries, cucumbers, peppers, squash, and strawberries. In Alachua County, near Gainesville, blueberries, bush beans, cucumbers, peppers, potatoes, and squash are the principal crops. In the north-central counties near Sanford, cabbage, carrots, celery, cucumbers, and radishes grow, plus the cooler weather crops—greens, escarole, lettuce, and spinach—that don't do well in subtropical South Florida. Nearby Zellwood, near Lake Apopka, is famous for sweet corn, while Webster produces eggplant along with cucumbers and peppers. Other crops grown in Florida are peanuts, corn, sugarcane, tobacco, sorghum, beets for sugar, and some wheat. These are called field crops, and the labor required is often difficult.

CHAPTER 2

Wages

Immokalee

Immokalee is a tiny farm town southwest of Lake Okeechobee and just northeast of Corkscrew Swamp Sanctuary, near Florida's Everglades region. The town grew up in the 1940s, when wartime agriculture expanded in South Florida. Much like Belle Glade, a few miles to the northeast, Immokalee has a large population of farmworkers. In the winter, when the main harvest season comes around, it swells with Haitian, Guatemalan, and Mexican workers. In the summer, the town is almost empty because most of the workers then travel north to the truck farms of New Jersey and Pennsylvania and the apple orchards of New England.

State Road 29 serves as Immokalee's main street on its way through town, before turning south to the Everglades. Most of the town's restaurants are Mexican, with signs in Spanish advertising frijoles and tacos, but there are a couple that specialize in "country cooking," serving up the standard bowls of black-eyed peas and greens, staples of the rural South. Down the side streets, slightly out of view from the main road, are shabby trailers and dilapidated cottages where farmworkers live during the winter months and the spring harvest season. African Americans, now year-round residents since most have settled out of the migrant stream, work in the packinghouses, while the Hispanic and Caribbean workers pick in the fields.

On a road winding out of town stands a Seminole bingo hall that seems strangely out of place in this rural town. When asked if he ever played bingo, Pedro, a Mexican farmworker who is active with the Coalition of Immokalee Workers, tossed his head back and gave a hardy laugh as he replied, "Diñero. No diñero."

Recently the small town has made national news several times because of its active farmworker community and the Coalition of Immokalee Workers. The Coalition is an organization of local workers, mostly Hispanic, Haitian, and Mayan, who perform farmwork and other low-paying jobs. In 1997, the group supported the tomato pickers who went on strike for higher pay. In 1999, several pickers with the Coalition went on a hunger strike for three weeks, attracting the attention of former president Jimmy Carter, who came and spoke with the fasting workers. The pickers were asking for a raise from 45 cents to 50 cents per bucket, or half-bushel.

Historically, farmworkers have received some of the lowest wages in the United States, and farmers may short even this meager pay. Until recently, bosses usually paid workers in cash, leaving no record to trace the hourly wage. Even though there are strict federal laws about minimum wages, these are difficult to apply to farmworkers. First of all, it is customary to pay these workers by the piece, such as by the half-bushel of tomatoes or cucumbers, or by the sack for citrus. The law requires farmworkers be paid a minimum wage of $5.15 per hour, but they must earn that through piecework. Therefore, at 45 cents per bucket, a worker must pick approximately one dozen buckets per hour to earn the minimum wage. If a worker cannot do that, then the boss may adjust the records so it appears that the worker spent less time in the fields; in this way both pay and hours appear to round out to the required wage. This gives workers an incentive to work harder, often exhausting themselves and working in poor health to get enough money for their needs.[1]

In October 1999, a survey by the Florida Department of Agriculture estimated that farmers hired 58,000 workers. The wages for hired field-

2.1. Workers wait in line with half-bushel buckets full of tomatoes. Each bucket weighs from twenty-five to thirty pounds. The crew leader dumps the tomatoes into wooden bins, quickly inspecting the condition and the ripeness to determine whether to give credit to the picker in the form of a poker chip, which the worker later redeems for wages. A bucket of first-picked cherry tomatoes earns a picker $2.00; the later picking earns only $1.25. Plum tomatoes earn 75 cents per bucket and salad tomatoes only 40 cents. The pay varies depending on how numerous the tomatoes are on the vine, whether they are first pick, and how many buckets a picker fills in an hour.

2.2. These men weigh their beans on a scale in a field near Belle Glade while someone records how much they will earn.

workers averaged $7.05 per hour, a rate that has remained constant for the last several years. The national average for hired field-workers in October 1999 was $7.31 per hour. Though this figure is competitive with service and entry-level jobs, it is an average of what the worker earns by the piece and the minimum wage salary, not a reflection of a standard pay rate.[2]

The average hours worked weekly by agricultural laborers in Florida during 1999 totaled just under forty hours in most cases. The state divides farmworkers into two categories: the 11,000 who work 149 days or fewer, and the 47,000 who work 150 days or more. In July, at the peak of Florida's midsummer heat, the number of workers drops to about 40,000 for those who work long-term, and to about 5,000 or less for the short-term workers. This is the time when farmworkers head north from Florida to the Northeast or Midwest to pick apples, cherries, and other summer crops.[3]

The average farmworker in America earns between $8,000 and $9,000 per year for labor, and those without steady work earn far less. Pickers earn a certain amount for each bushel, box, or sack filled, and each product demands a different price. When the Immokalee's tomato pickers first went on strike in 1997, Mike Stuart, president of the Florida Fruit and Vegetable Association, maintained that the area's farmworkers earned $8 per hour, or about $16,000 annually. Farmworkers countered that the estimate was more than anyone could realistically earn because it was calculated on a forty-hour work week, which most workers seldom see.[4]

Though the figure Stuart mentioned might seem high, a 1989 California study shows just how the money usually breaks down. In four farmworker communities in California, 65 percent of the persons in farmworker households worked seasonal or temporary farm jobs. The gross annual household income averaged $15,200, but, considering that the average household has six to seven members supported by only two to three workers, these people actually lived below the defined poverty level of $20,000.[5]

During the 1990s, the actual wages for farmworkers became proportionately less than in earlier decades. With the strong economy, the salary of the average nonagricultural worker rose about 32 percent, but for farmworkers, the increase was only 18 percent, from $5.84 to $6.18. Adjusting this figure for inflation shows that these workers lost 11 percent of their buying power, so that relative to the pay for other jobs, farmworkers' salaries went down.[6]

The low wages affect every aspect of farmworkers' lives. Most families can only afford minimal housing, usually trailers, and the landlord may charge by the person rather than just by the month. Some workers rent the grower's trailers at prices that can go as high as $400 per week if the grower charges by the person. With so much required for rent, money for food is scarce, and malnutrition is a genuine worry. Farmworker advocates say growers' estimates of salaries reaching $16,000 per worker annually might be accurate for a few days at the season's peak, but such earnings do not continue year-round. During the summer months, when there is little work in Florida, workers often use what little money they have to follow the growing season north. During the torrid Florida summers when few things grow, those who choose to stay may find sporadic odd jobs, but the financial situation is always tenuous.[7]

Most advocates consider the inadequate wage the basic problem in farmworkers' lives. Florida is a right-to-work state where unions hold little bargaining power. Though there are alliances and coalitions, they carry little weight with the large growers, who may dismiss employees who cause trouble. With the increasing influx of undocumented and unaccompanied male workers streaming in from Mexico, union organizers in Florida seldom find people willing to protest. Quincy, in the Florida panhandle, is the exception. In the late 1990s, the mushroom workers joined the UFW.

"These men expect to work for just a couple of years and then they want to get into construction or service jobs," says Gregory Schell of the Migrant Farmworker Justice Project. "They are undocumented workers. They may show you some papers, but they're not real. These men are the ones doing the most migrating just now, and the farmers are happy with them. With the increase of corporate farms, these men provide easy,

2.3. Ramona Zarate sells tacos from her car trunk at the edge of a field to supplement her own fieldwork income. She provides homemade food for the field workers, who often live in barracks-style housing with inadequate cooking facilities.

2.4. Lucas Benetiz (*left*), spokesman for the Coalition of Immokalee Workers, chats with a hunger striker and another farmworker at the coalition's office in downtown Immokalee. The workers struck against the tomato growers, asking for better wages. The coalition has a multinational membership, so the sign on the front window is written in three languages for the Haitian, Mexican, and Guatemalan workers. Here, an interior mural depicts workers with arms raised in solidarity in front of a mythic figure.

cheap labor; most of them won't cause trouble because they are undocumented; and they are far easier to house. They put these guys eight to ten in a room, packed in dormitory-style. They no longer have to worry about unproductive family members, as they did with the extended Mexican families who traveled with Grandma and underage children."[8]

Lucas Benitez

Lucas Benitez is originally from Guerrero, Mexico, but Immokalee is his present home. He began picking oranges at the age of sixteen, and at twenty-three he was recognized as a leader in seeking rights for farmworkers. He is a member of the Coalition of Immokalee Workers in the tiny farm town. In 1998, he was honored by the Catholic Bishop's Conference, receiving the Cardinal Bernardin Award for New Leadership, which is given annually to one outstanding person under thirty. In 1997, Benitez acted as a spokesman for the hunger strikers during their month-long protest for higher wages, although he did not participate in the hunger strike himself. When former president Jimmy Carter came to Immokalee to ask the hunger strikers to stop their fast, the workers stopped out of respect for Carter, but they refused his help, says Benitez. They only wanted to have a dialogue with the growers, Benitez explains, and he offers ideas on what could be done to improve conditions for farmworkers in the future.

"It's a fundamental problem in Immokalee that the workers get no respect from the growers," Benitez explains. "The growers think they have peons, not employees. To find solutions to other problems, we must break down that barrier. Stemming from that [lack of respect] is low wages. We've raised them some, but they are still quite low. We should be making 73 cents a bucket for tomatoes, but we only get 45 cents or 50 cents from Gargiulo. Twenty years ago, we made 35 cents to 40 cents per bucket.

"The figure of 73 cents per bucket is based on what workers should be earning in 1999 as a cost of living increase from twenty years ago. The only grower who has given even a small increase is the Gargiulo com-

2.5. This woman is picking tomatoes grown in long rows. She will have to pick about ten buckets per hour to make $40 in an eight-hour day.

pany, a large tomato grower who increased workers' pay by 5 cents per bucket. It seems insignificant, but it does add up for pickers. As long as this problem of low wages persists, it is impossible for their families to get ahead financially.

"The growers provide no housing in Immokalee; it is all private," Benetiz explains. "There are no company stores here. That practice is nearly gone everywhere, but all the stores here charge higher prices because they have a captive clientele. Winn-Dixie is the only big store, and the other smaller stores are owned by Mexicans and Arabs."

When asked if the Mexican owners have any compassion for their own countrymen, Benitez responds, "They are businessmen—they only care about making money."

Though most of the company stores are gone, there are still some cases of debt peonage. This is the notorious system in which a boss recruits workers and then keeps them in permanent debt by overcharging them for transportation, room and board, food, liquor, drugs, and sometimes even prostitutes.

"There were twenty-two convictions between 1978 and 1986 for debt peonage. These prosecutions in the early eighties, about peonage, were against Florida-based African American crew leaders who got people out of [homeless] missions," says Greg Schell.[9]

Benitez discussed another disturbing problem, that of farmworkers held in slavery. Though some workers come to this country legally as temporary agricultural workers, there is no guarantee the recruiters are honest crew bosses. There have been a number of instances in which workers were confined to their quarters when not working in the fields, making them modern-day slaves. In several other cases, immigrant women were held and forced to work as prostitutes for the farmworkers until they paid the coyote his fee for smuggling them into the United States. Usually the smugglers lure young women with the promise of a lucrative job as a maid or a nanny.

"We still have slavery cases," Benitez continues. "Recently in April of this year we gathered information on a crew leader who held male workers in slavery. Five workers escaped from the remote labor camp near

Palm Beach County and reported the abuses to the Coalition. We contacted the Department of Justice and with their help we rescued twenty-five people. The youngest was fourteen. Most of the workers were of all ages, but there were several who were underage. We had been suspicious. One of our board members reported it. We've been educating people to recognize this problem."

Benitez explains how to recognize farmworkers held in slavery.

"When people are not free to leave on their own, then they are being held against their will," he says. "When these people wanted to leave, the crew leader would not let them."[10]

Immokalee and Benitez were back in the news in 2000 with a renewed protest over wages. This time the Coalition marched 250 miles from Fort Myers to Orlando to ask for better pay. The Coalition's membership is over 1,000 Mexican, Guatemalan, and Haitian workers. They are demanding to sit down at the table and negotiate a wage increase with Immokalee's largest growers. Their demand is an increase from 40 cents or 45 cents a bucket—which is half a bushel—to 75 cents a bucket. Protesters say that would give them a living wage.[11]

Meanwhile, the Coalition paid workers who registered with the protest about $7 daily for living expenses while they were on strike. Organizers collected over 1,700 signatures on petitions sent to growers and asked to meet with them. The Florida Fruit and Vegetable Association, a growers' group, said they do not consider the Coalition representative of their workers, and they continue to refuse to meet with Coalition leaders. Growers said pickers working for them earn ten times what they earn in Mexico, and if they raise wages, they would be unable to compete with Mexican produce.[12]

Greg Asbed

Greg Asbed is the coordinator for the Coalition of Immokalee Workers and works alongside Benitez lobbying for better wages for farmworkers, especially the local tomato pickers. He is busy and energetic, often traveling to speak for farmworkers. He talks about the conditions of the

2.6. Lucas Benetiz leans against the front window of the office of the Coalition of Immokalee Workers as the worker in the middle tells about former president Jimmy Carter's coming to Immokalee to visit the tomato pickers on the hunger strike.

workers and the continuing quest to have a discussion with the tomato growers.

Though Asbed is not thrilled with recent proposals by Florida Senator Bob Graham to allow immigrants to become citizens after five years of working in the United States, he does think the senator is genuinely interested in improving conditions for farmworkers.

"We've got a lot of political momentum right now with Graham," says Asbed. "He's interested and he really wants to do something. So far, the growers have not responded to our requests to meet with them. We sent another letter to [Governor Jeb] Bush, asking him to come. Another thing we've done is talk to the major commercial buyers of tomatoes, like Taco Bell, to bring the workers to their attention."

In April 2001, the Coalition of Immokalee Workers kicked off the official "Taco Bell Boycott and Truth Tour" with a big party and protest at a Taco Bell in downtown Orlando. Their slogan is "*Yo No Quiero Taco Bell!*" (I Don't Want Taco Bell!), a parody of the ads the company produced on television featuring the coercive chihuahua. According to a statement issued by the Coalition, farmworkers who pick tomatoes for the Immokalee-based Six L's Packing Co., Inc., one of the nation's largest tomato producers and a contractor for Taco Bell, are paid 40 cents for every 32-pound bucket they pick. That is the same per bucket rate paid in 1978. At that rate, workers must pick and haul two tons of tomatoes to make $50 in a day. The coalition maintains that workers picking for Six L's are denied the rights to organize and to receive overtime pay, and they receive no health insurance, no sick leave, no paid holidays, no paid vacation, and no pension.[13]

Asbed also mentions another problem the Coalition is concerned with—that of "pinhookers." *Pinhooking* is the term used to describe farmworkers picking from the fields and selling the produce themselves. It is usually either what is left after the field is picked several times or excess from the packing plants. For farmworkers, it is an extra source of income.

"The growers don't like pinhookers, because they want to sell every-

2.7. Cabbage is labor intensive. Workers use knives to cut the stems, and they are paid by the head. Here the cabbage is put on this conveyor belt and hauled up to the tractor that pulls a bin for workers to fill.

2.8. Workers follow the crops from the planting to the harvest, often even packing it for sale. These men are packing crates of zucchini for market.

thing themselves," says Asbed. "Now the growers want crews to box up the tomatoes in the fields to prevent pinhooking. They are asking everyone to bring workers to the fields to pick, so there are many people who are inexperienced at running crews who are becoming small-scale crew bosses. This makes it hard on the farmworkers if they work for a crew boss who doesn't know what he's doing. These small bosses often work as pinhookers."

Asbed says that most of Immokalee's workers go north in the summer, while some stay and find work picking saw palmetto berries that they can sell to buyers from pharmaceutical houses.

"Saw palmetto berries are good in the summer. It's a good business, but there are rattlesnakes around because the palmetto trees are out in the wild. Nobody has saw palmetto farms. The berries are priced by the pound, and buyers come to Immokalee and buy them on the street."

Laura Germino

Laura Germino holds a master's degree in community development and works with Florida Rural Legal Services in Immokalee. She is married to Greg Asbed, who works with Lucas Benitez and the Coalition of Immokalee Workers. Germino describes Immokalee as "unique in its consciousness raising for social change," asserting that her own program with Florida Rural Legal Services is committed to educating farmworkers so that change can occur through the community itself. To achieve this, the program develops ties to the community and offers technical and legal support to the workers to increase their awareness of their labor rights.

As an example of this community work, Germino discusses a slavery case that her program had investigated for five years, an investigation that resulted in Miguel Flores's arrest and fifteen-year federal prison sentence: "Out of the Flores case, we forged a partnership between the community groups and the federal government." Describing Florida Rural Legal Services as playing "the helping role of the intermediary," Germino says that as the program teaches farmworkers how to negotiate for their labor rights, these workers in turn teach these skills to other farmworkers: "When they go north to pick, they teach other farmworkers, so we don't have to be there. The workers become involved themselves."[14]

Other cases of slavery and enforced prostitution have occurred around the nation in recent years. Justice Department officials say they have prosecuted ten such cases involving 150 victims over the last three years. Groups that help abused workers assert that the crime is widespread but difficult to uncover because it is well hidden. Language and cultural barriers often isolate the victims. In April 1998, officers busted a large prostitution ring that allegedly recruited women in Mexico, brought them to Florida, and held them in several cities, forcing them into prostitution. The smugglers charged the women $2,000 each to bring them into the country, then forced the women to work off the debt as prostitutes for Mexican migrant workers. Sixteen Mexican men faced federal charges in Fort Pierce for bringing these women into the country.[15]

Though most of the growers in Immokalee refused to listen to the Coalition when they raised their voices, one company did. Gargiulo Inc., the company that increased the workers' piece rate for tomatoes by 5 cents is also one of California's largest strawberry growers and a subsidiary of Monsanto. In 1998, they agreed to pay $575,000 in back pay to hundreds of California workers who were forced to work without pay in the early mornings before their shifts officially began. The AFL-CIO bankrolled a class-action suit by the United Farmworkers (UFW), the union founded by Cesar Chavez in 1960, and it was a success.[16]

In March 1998, author and feminist leader Gloria Steinem joined 1,000 activists in a New York City march to protest the terrible working conditions in the California strawberry fields. Chanting "*Se puede,*" the Spanish slogan for "It's possible" popularized by Cesar Chavez, the group marched down Broadway past stores that sell strawberries picked by women who do not earn enough to feed their families. Growers had fired California workers who had tried to unionize in the strawberry fields, according to United Farmworkers. Dolores Huerta, who cofounded UFW

2.9. Picking strawberries can be prickly work, so gloves protect this woman's hands from the red juice as she harvests the fruit near Dade City.

with Chavez, said that California's strawberry workers earn about $8,000 annually and have no health insurance or other benefits.[17]

Mushroom farms are another battleground for workers demanding better wages and working conditions. Quincy Farms, in the small Florida panhandle town of Quincy, supplies twenty-five million pounds of mushrooms annually to seven southeastern states. In 1998, workers filed suit against Quincy Farm's parent company, Pennsylvania mushroom giant Sylvan Inc., challenging their unpleasant working conditions. Quincy Farms workers reported they had to ask permission to go to the bathroom, and that they lacked needed ladders to climb between the large mushroom beds that stood nearly twelve feet high over damp concrete floors. In March 1996, when eighty-six workers were fired for a protest walkout, only twenty-four were rehired. Six mushroom workers filed the class-action suit against Quincy Farms on behalf of the fired workers. Litigation continued in the suit until January 1999, when the company reinstated the remaining workers fired in 1996.[18]

Then, in July 1999, 450 workers at Quincy Farms signed a union contract with United Farmworkers, making them the only unionized farmworkers in right-to-work Florida. The eighteen-month contract applied to all of Quincy Farm's workers, whether or not they signed. Two hundred and fifty employees received raises of 25 cents to 50 cents an hour, bringing their hourly wage to about $5.75. Some of the workers qualified for benefits, and all workers were guaranteed health insurance and a pension plan, plus paid vacations. Even workers who were paid 16 cents per pound for picking mushrooms received profit sharing, medical insurance, and a union-sponsored pension plan. These are extras nearly unheard of in the world of farmworkers.[19]

Filling Jobs

Currently there are two workers, three by some estimates, for each available job in the fields. That means farmers can offer a very low wage and still find someone willing to work. The H-2A system—an updated version of the old *bracero* program of the 1940s and 1950s—allows growers to bring in foreign workers legally if enough local workers cannot be found to fill the jobs. In 1996, the U.S. Department of Immigration instituted more stringent requirements that kept many foreign workers out of the country. Since then, many farmers have complained bitterly that their harvests are rotting in the fields for lack of labor, particularly in California, where the flow of Mexican labor slowed with the new laws. Farm labor leaders claim farmers want to hire foreign workers because they are easier to manage. Because these foreign workers, mostly Mexicans, can be brought in for only the period they are needed, growers find them preferable to workers who live in this country. If they are unruly, or if they protest their wages or treatment, they are quickly deported.[20]

In 1991, a landmark case affecting all farmworkers was decided in United States District Court for the Central District of Illinois. In this action, a group of migrant farmworkers sued the owners of a farm in Kankakee County, Illinois, where they worked seasonally from 1983 to 1988. Three of the four counts were class actions, but the individual plaintiffs brought the fourth count. The workers sued under the Migrant and Seasonal Agricultural Worker Protection Act (AWPA), proving that there are working laws to protect their rights.[21]

The workers claimed the farm violated their rights in a variety of ways: They accused the employers of keeping poor work records and providing inadequate housing. They charged that the growers did not comply with federal or state law in withholding certain portions of their hourly paychecks. The workers also claimed that their employers failed to have required insurance on certain farm equipment and failed to provide them required information about the terms of their employment. In count one, the workers filed under AWPA, claiming that the growers did not keep pay records as required by law. The defendants responded that since theirs was a family farm, they were exempt from the requirements of AWPA. However, prior court rulings defined farm labor contracting as "recruiting, soliciting, hiring, employing, or transporting any migrant or seasonal agricultural worker." Where evidence exists that a nonfamily member performed this hiring, the exemption does not apply. The court found that nonfamily members transported workers, that the employers

2.10. Workers gather inside a parking area where a bus will take them to the fields, usually an hour or so before dawn. In California, some workers won a settlement because bosses made them start working before their shifts began.

kept no records, and that vehicles used to transport the workers carried no insurance. The growers also neglected to pay Social Security into an account for the workers. Much to the defendants' dismay, the court decided in favor of the farmworkers, mandating a jury trial that resulted in the award of back pay for the workers. These are small victories, but victories that illustrate how easily farm owners and crew bosses can take advantage of foreign workers.[22]

Foreign Workers

Prior to 1986, the program of importing foreign workers was called the H-2 program, but since the passage of the Immigration Reform and Control Act of 1986, it is called the H-2A program. There is little difference between the two programs. The H-2A program allows farmers to recruit workers from outside the country if they can prove they cannot find local workers to do the job. This program is supported by Congress, but farmworker advocates oppose it.

Fernando Cuevas Jr. runs the Farm Labor Organizing Committee (FLOC) for Florida, keeping track of all requests that come into the Florida Department of Labor asking permission to recruit foreign workers. His father, Fernando Cuevas Sr., is the national vice president of FLOC. Both men are deeply involved in farmworkers' rights. While his father is busy with his national duties, Cuevas Jr. keeps an eye on Florida's problems.

"I try to find local workers to do the job, because there are many here out of work," says Cuevas Jr. "We request the records of who wants to bring in workers, and if we find local workers, they can't bring in the H2s."

Though Cuevas Jr. says growers guarantee H-2A workers $6.77 per hour for cutting, planting, and weeding the fields, bosses often touch up the records to indicate that the worker actually made $6.77 an hour, when in fact he probably worked much longer to earn that amount. Again, growers find that the H-2A workers, who are mostly single men, are less likely to complain or to involve themselves in organizing efforts.[23]

2.11. This citrus picker dumps his bag that weighs about 90 pounds into a bin that will later be tallied to determine his pay. In 1985, citrus workers earned about $148 for picking 19,800 pounds of fruit.

As workers become more educated, they are able to find agencies and advocates willing to take their sides and demand fair wages. Though there are certainly cases of cheated workers, today many more are willing to come forward than before. Small victories and increased public awareness of the unfair wage practices bring many growers to task, forcing them to pay workers a legal wage. Though the 1938 Fair Labor Standards Act excluded farm laborers from the protections other workers received, the 1966 revision included them, granting them the minimum wage and Social Security benefits they were previously denied. Like other seasonal workers, farmworkers can now benefit from unemployment insurance, but they were not allowed to collect unemployment benefits until 1976. Many of the stories of abuse reported in the media are sensational, meant to shock the sympathetic reader. In reality, most workers are too ambitious and too independent to stay with an abusive boss. Like other immigrants, these workers want to advance, to save money for their families, and to improve their standard of living.[24]

More than forty years ago in Belle Glade, the study of the citizens' committee recorded that a citrus picker received 17 cents per box for oranges. Picking at the average of 75 to 100 boxes a day, the worker's earnings would total from $12.75 to $17.00 daily, to make a six-day work week pay between $75.00 and $100.00. According to the report, crews that worked the full six days owned homes and automobiles, but some crews averaged only about three and a half days and earned only $45 a week. In 1985, one citrus picker's pay stub showed a total of $148 for harvesting 19,800 pounds of fruit. From that figure, a deduction of $55 covered rent, lights, and gas, leaving the worker with $82 for a week's wages. Today's wages have increased, but not proportionately to those for other occupations. Farmwork remains one of the lowest-paying occupations, with workers fighting for every penny.[25]

CHAPTER 3

Housing

In Dade City in Pasco County, there is a small community of wooden houses nestled away from the main part of town, where most of the residents speak only Spanish. Here the roads are unpaved, and they pool with muddy water when it rains. Local Dade City residents call the area "Tommytown." These are the homes of the farmworkers, mostly Mexican, who pick the strawberries that are a major crop in eastern Pasco and Hillsborough counties and make nearby Plant City the "Strawberry Capital of the World." The wooden bungalows and trailers in the neighborhood are typical of farmworker communities in Florida and around the nation.

Housing is a major problem for migrant farmworkers. Torn screens, windows taped up to hold broken glass together, and trash in the yards are all common sights in the run-down communities where most farmworkers live. Space and other amenities most Americans take for granted are scarce in these minimal dwellings. Most of these labor camps are nearly invisible, usually situated far away from main roads but near the fields where the workers pick the fruits or vegetables. Because of low wages, workers usually cannot afford a place large enough for the entire family, and they often live in cramped, very crowded conditions, with several families sharing one residence. If they live in housing provided by the farmer, it may be inadequate, and the rents are high because farm-

3.1. An extended family lived high on this hill in Central Florida overlooking the fields they harvested. The rutted quarter-mile road leading from the highway became nearly impassable after a heavy rain. Screen was stapled over the windows to keep out mosquitoes.

3.2. Families often do laundry by hand, hanging it out to dry. Dozens of socks, jeans, and other work clothes hang in this fence near a trailer. Most housing for farmworkers does not have laundry facilities.

workers have few other options. Farmers are not required to provide workers any free housing in Florida, so landlords sometimes take advantage of this. If farmworker families live in state-subsidized housing, they cannot keep their apartments when they go north for a few months, even if they pay the rent.[1] This means that families cannot have any furniture or permanent household possessions. Migrant farmworkers can own only what they can carry with them.

Margarita Romo of Farmworkers Self-Help works hard to improve housing in her Dade City enclave. Many of the people she works with are Mexicans who have settled out of the migrant stream, but she still sees some workers who come through the area with their only possessions in a backpack and just stay for the night at the small mission. Romo says that there are migrant camps in Dade City, but that crew bosses own them, not farmers. Farmers have passed the problem on to the crew chiefs, but exploitation continues.[2]

Housing is much the same in farming communities around the state. In Immokalee, where the entire economy is based on vegetable farming, especially tomatoes, the housing is typical. Just off the main highway that runs through the center of town are the homes of the farmworkers. Some are small, run-down shacks, barely painted and situated in rows on a small lot. Some are trailers, crowded cheek by jowl in an area that allows little privacy. Broken-down cars and laundry blowing in the breeze are common sights. Unkempt dogs lurk under porches, waiting for a chance to bark. There is a sharp contrast between the homes of the workers and those of the crew bosses and growers, who live in conspicuously lavish houses nearby.

Ignacio Romero runs the office of the Farmworker Association of Florida in the farming community of Immokalee, where he directs a program that assists farmworkers in buying their own trailers. Through Romero's program, farmworkers can purchase a home for far less than the monthly rental charged by most landlords, who frequently gouge the tenants for houses or trailers in dismal conditions.

3.3. This man, who was hurt falling from a ladder in an orange grove in Central Florida, stands outside his trailer at the labor camp where he is staying. This trailer is typical of those available for workers, especially those who migrate from harvest to harvest. Even pickers who stay in one area can usually only afford to rent trailers.

"We have bought 92 trailers and sold them to farmworkers for $1,000," says Romero. "It saves them about $800 a month, because before they were paying $800 a month for less than good trailers."

Romero says the program is ongoing, and he hopes to continue to subsidize these trailers in the future as funds become available. These mobile homes allow the workers to settle out and become more stable. Even if they continue to migrate for part of the year, it gives them a place to call home.[3]

Another program in the small Collier County town of Bonita Springs has successfully placed farmworkers in rental homes that are decent and affordable. The farmworkers' village, known as Pueblo Bonito, is the special project of Reverend Don Franck and his wife, Gerry, who spent five years bringing it to fruition. When the Francks began to promote the idea of Pueblo Bonito, they faced strong opposition to the plan from local residents who feared a run-down, crime-ridden area typical of much government-subsidized housing. But in 1995, floods drove nearly 1,800 residents of Lee County from their homes. Many were farmworkers who lived in low-lying areas, who found themselves homeless and took shelter at Estero High School in Bonita Springs for a few weeks. The floods brought in federal money, and soon state and county money followed. In June 1999, with the help of the Francks, eighty farmworker families qualified for the new housing, and ninety names were on the waiting list for another seventy-three units to be ready soon.[4]

Though the community was divided over the project in the beginning, Justice Department civil rights attorney Marta Campos persuaded county commissioners to rezone the land for Pueblo Bonito under the fair housing practices. Campos became involved after county commissioners refused to rezone the land for Pueblo Bonito, but she dropped all charges after Lee County approved a modified plan. Supporters say this project will not be the last of its kind. Thomas Pierce, acting division director for Housing and Community Development for the Florida Department of Community Affairs, said that the federal government allocated $10 million for nationwide farmworker housing in 1999, and that he expects to use some of the money to build 450 homes in Florida.[5]

3.4. Here a woman cooks tortillas on a griddle in the small kitchen of a trailer. She is still wearing the clothes she wore in the fields as she prepares supper as well as the next day's meals. Small electric appliances, such as frying pans, toaster ovens, and perhaps a hot plate, are often all that is available for cooking.

Evan Jorn

Sometimes local communities decide that farmworker housing is not up to code, and local authorities will try to force landlords or tenants to clean it up. By enforcing violations, they frequently frighten the workers, who are financially unable to fix their surroundings. Such a sweep happened in mid-1999 in southern Hillsborough County near the rural farming communities of Wimauma and Balm. The nearby Beth-El Mission, a Presbyterian outreach mission, was instrumental in helping farmworkers respond to these charges of violations. Mission director Evan Jorn noted that the code enforcement unit was completely out of line in the way it handled the issue.[6]

"They [code enforcement unit] went to the homes—many of them are mobile homes—and they put the green stickers on them. These stickers say this dwelling is unfit for human habitation, and that it is unlawful to enter the dwelling under any circumstance. This green sticker was used in the first instance, but that sticker shouldn't be used until a lengthy process is completed. And according to the sheriff, it's not illegal to be in the building.

"Condemning a dwelling usually takes time. Three people must make the decision to condemn the place. The residents have lots of time to fix these violations, and the green sticker should not be used unless the problems are not fixed after several notices," he said.

"The green stickers were in English, so some of the workers who speak no English could not even understand them. A number of the people were frightened to enter their homes and deeply upset. Most did not know how they would be able to fix their homes at all.

"Ironically, it was good because many people called and donated money," said Jorn. "Sun City Center [a nearby community] sent funds. We formed the Balm-Wimauma Affordable Housing Partnership, a community partnership between the Good Samaritan Mission, Bay Area Legal Services, Sun City Center, and SunTrust Bank. The funds are in SunTrust Bank, and we use the money to make grants or guarantee loans so people can fix up their homes.

3.5. These nearly invisible trailers stood at the edge of fields where workers labored daily. Three families and a number of single men lived in these five mobile homes. A dirt road was the only way to reach them. (Photo by Nano Riley)

3.6. These tiny one-room houses in a labor camp near Lake Apopka are only about ten by twenty feet. The only plumbing is a kitchen sink and a small bathroom to the side with a toilet. There is no shower. These cabins typically house several men or a family of farmworkers, who usually sleep on the floor in sleeping bags or on mattresses. A grower owns these cabins, but they are vacant now because the nearby farmland was part of the Lake Apopka restoration project discussed in chapter 6.

"If a farmworker gets a loan from SunTrust, it is guaranteed by the fund, but no money is withdrawn from the fund unless the person is unable to repay the bank loan. It gets people into the credit world. If they repay the first loan, then the next time they need $2,000, they have the credit they need to borrow from the bank. It's a community banking system that improves their credit," said Jorn. "We also organized a community cleanup, and we went in and removed fifty tons of trash in six days. Do you know how difficult it is to organize one of those cleanups? We did it, with funds from the Children's Board used to get government help."[7]

The problem in Balm is mirrored around the country in poor farmworker communities. In February 2000, the United Farmworkers in Farmersville, California, issued a plea to help nine farmworker families, including twenty-seven children, in jeopardy of losing their homes in the middle of winter. The families faced imminent eviction from their run-down homes, which were owned by slumlords and condemned by the city. Some of the nine families in Farmersville had no indoor toilets or washing facilities. They shared an outdoor wooden outhouse with a toilet and shower. The landlord charged $550 per month for a tiny two-room unit rented to three people.[8]

Marvin Brown and Jay Sizemore

Marvin Brown and Jay Sizemore are strawberry farmers in eastern Hillsborough County who want to do something positive for farmworkers. In 1990, Brown built a twenty-unit housing development in Dover next to his strawberry farm. Now, he and Sizemore are partners in the strawberry business at JayMar Farms, and they have plans to build a new seventy-two unit development on thirteen acres in Balm off of County Road 672. The citizens of Balm protested the project, basically because they did not want housing for farmworkers in their midst. The county zoning master approved a plan for the project in October 1999, but the Balm Civic Association appealed. Brown and Sizemore felt confident about the outcome and proceeded with the beginning engineering for their plan so that they would have housing ready for the fall planting season. Both men say it is just good business, because if they offer good, affordable housing, they will attract a better workforce.[9]

Sizemore says that most of the workers on the JayMar strawberry farms are seasonal, traveling up the East Coast as the work here dwindles in the summer. However, there are some families who stay year-round except for a short time in the summer, when they go home to visit relatives in Mexico. According to Sizemore, he and Brown have made lots of concessions in order to appease the Balm residents.

"Basically they just don't want that many people living near them, and they think the housing will get run down," Sizemore says. "They [the Balm residents] just aren't understanding. They wanted us to screen the houses so they won't be seen from the street; they wanted us to add storage sheds so the workers wouldn't keep things in the yard; and they wanted a place for them to change their oil.

"We'd just as soon they stayed year-round," he says. "A lot of these workers want to get out of the migrant deal 'cause it messes up the kids in school. We have crops that last 'til June, so the kids can finish school."

Sizemore's partner, Marvin Brown, a native of Hillsborough County, was named the 1998 Agriculturalist of the Year by the Plant City Chamber of Commerce. Brown has worked closely with farmworkers for the twenty-two years he has been a strawberry farmer. Some of his employees have been with him for more than twenty years. The duplexes Brown built in Dover have two and three bedrooms, and he rents them to his workers for $150 per month. The new units that he and Sizemore are building in Balm are three- and four-bedroom units, and the rent will be about $200 to $250 per month. Brown explained why he wants to provide decent housing for his workers.

"I've worked with these people for years," says Brown. "I hunt with them, fish with them, eat at their homes, and go to their parties. We need to do the right thing. We want year-round workers, and we prefer families, but there are a few single men. We have berries in the ground about nine months out of the year, so there's work for them.

"We get the usual complaints," Brown says, referring to the Balm resi-

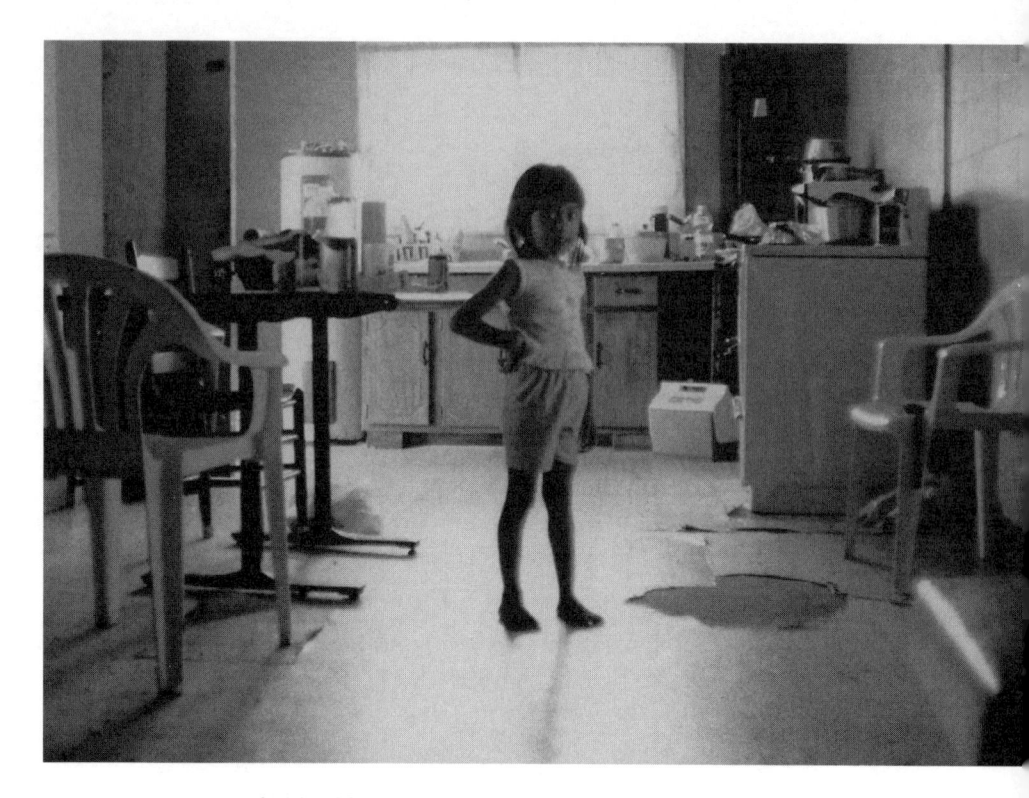

3.7. A family of five lived in this house with torn linoleum, bent chairs, and a broken refrigerator in the kitchen. Sometimes as many as five more relatives joined them. Lacking refrigeration, coolers served to keep food, which spoiled quickly.

dents who want to curb the project. "'Things will come up missing,' they say, or 'there will be trouble in the neighborhood.' They don't know these workers like I do. Most of them are family-oriented and they work hard. They don't want to cause trouble.

"These units are just for our workers," explains Brown. "They have to keep them up and they must abide by the rules and regulations we establish. They are responsible for any damage. Generally when you put someone in a nice house, they have an incentive to keep it up. I just repainted the duplexes in Dover inside and out for the first time, and they were clean. There was no graffiti on the outside."[10]

Brown and Sizemore both say they want to "do the right thing" and that the housing is a win-win situation.

"It's a good thing all around," says Sizemore. "Many people think somebody's got to get burned, but that's not the case here. It's good for us to have a good work force."

Brown says that the housing provided by the strawberry industry in eastern Hillsborough County is the best in the state.

"We won't make any profit on our housing units, but it's the right thing to do," says Brown. "We need to treat all people the way we want to be treated. I have an impeccable reputation for the way I treat my workers."[11]

Other projects slated to improve farmworker communities are on the books as the twenty-first century begins. In the dilapidated area of Dade City called Tommytown, Pasco County government allocated $6 million to renovate the neighborhood, beginning with street paving and installing water and sewer lines. The county is also offering a financial incentive for people who want to buy or repair homes. They are eligible for 80 percent financing, interest-free, from the county.[12]

Migrant farm labor camps are governed by the Department of Health in Florida, and the statutes and regulations are administered by the county health department. To establish a labor camp for workers, the owner of the facility must file a permit with the state and must prove that it provides adequate personal hygiene facilities, lighting, and sewage and garbage disposals. The permits are supposed to be posted all the time, but

3.8. A ringer washer assists this woman in her daily laundry chores. She will hang the clean clothes out to dry on a line.

they are not always placed where they are easily visible. If these camps do not maintain certain standards, they can be shut down until they are brought up to code. They are also subject to inspection at any time by the county health inspectors, who can issue citations if there are violations. Advocates say that these inspections of farmworker housing are usually only conducted if there's a complaint, or if a resident of the camp receives welfare benefits.[13]

Because farmworkers often live in cramped conditions, safety is a concern. In 1998, Felipe Avelino, twenty-six, lost his wife, Maria, twenty-one, and three children, ages five to eleven, in a fire at their home in Dover. Authorities were unsure of the exact cause but said the fire began in the living room. It was just before Christmas, and the community reached out to Avelino. They purchased a ticket for him and donated enough money to fly the bodies of his family back to their home in Guerrero State, in southwestern Mexico, to be buried.[14]

Farmworker housing is much the same around Florida and in farm communities nationally. Because they are so transitory, most farmworkers do not complain about their homes. They usually spend little time at home except to sleep, especially if they are moving with the crops. They go to the fields at dawn and often work in the packinghouses until midnight. Pickers who have settled out often own their own homes or trailers, but for those who travel, little changes. They still live two and three families to a one-family home, and the single men cram in dormitory-style, trying to save as much as they can to send back to their families in their homeland.

CHAPTER 4

Education

Farmworker families have a difficult time escaping the cycle of uncertain work and enduring poverty. With education, they have a better chance, but education may be frustrating for these children for several reasons. If they are living in a migrant family, they must follow the harvests with their parents, so they are constantly changing schools. In their close-knit communities, Spanish is their first language, and they often speak English only at school. At home, these children become translators for older family members who may be illiterate and unable to help with schoolwork. By the time they reach adolescence, many of these children drop out of school and follow their parents into the fields. They do it because farmwork is all they know, or because they may want to help shore up the family's meager income.

In spite of the language barriers and self-esteem problems, there is help for those determined to succeed. During the 1960s, President Lyndon Johnson introduced social reform programs to address the problems of America's underprivileged. Title I of the 1965 Elementary and Secondary Education Act began as a program to offer children of poor families an equal education. In 1990, an amendment to Title I passed into law offering special programs designed to prevent dropouts. The

Migrant Education Program receives money under this provision and operates in areas with a high population of migrant children.[1] Though these programs have critics, most educators and advocates agree that they are beneficial to these underprivileged children, giving them the opportunity to leave behind the backbreaking labor of the fields.

Luana Peres and Juanita Cannon

In Hillsborough County, Luana Peres is the director of the Migrant Education Program funded by Title I. This program is assessed every five years and modified to fit current trends. Peres, who laughingly refers to herself as "Mother Migrant," has worked with the migrant program for twenty-six years, and she calls it is truly a rewarding job.

"It is a great program. You feel good knowing that you are doing good," says Peres. "The families are appreciative of everything. As society goes on, many people feel caught up in a system they can do nothing about, or they take things for granted, such as certain help programs, but we don't find that with the migrants. They are truly happy to receive help.

"A person who does not travel to do farmwork for three years is considered a former migrant and is no longer eligible for state or federal aid. It used to be they were counted as migrants for six years even if they didn't travel, but now it is only three years [without traveling for work] before they are considered former migrants," Peres says. "Many farmworkers will settle for about three years and then go back on the road for a time in order to keep their status as migrants and qualify for migrant services. Sometimes they return to the migrant stream because of a family breakup or the loss of a steady job. But as long as the family migrates for even a few months in the summer, they are considered migrants.

"Ninety-nine percent of the families in our program in Hillsborough are Mexican," Peres says. "They are more stable here in Hillsborough than in many other places. Recently they seem to be coming back earlier

4.1. At the Beth-El Mission in Wimauma, women study English using a computer program. Not only do they learn English, but they also gain some proficiency on the computer. One woman here said that she wanted to get a job at Kmart, but that she continues to work in the fields while she learns because it is all she knows.

and leaving later. They are coming back to the same area, for different reasons—maybe the weather or job availability. They travel more in the summer."[2]

Peres's assistant, Juanita Cannon, deals directly with the Migrant Education Program in Hillsborough County, providing education and services primarily in the rural southern part of the county. Even though there are fewer migrating families, it is still difficult for the children of this closed society to integrate and be successful in an ordinary school setting, Cannon explains.

"There are no specific counseling programs in the elementary schools. But the secondary schools in Hillsborough County with high farmworker populations have counselors to work with the children of the Mexican workers who live in the area," Cannon says. "Some families are still traveling, but many of the families who have been coming to the Dover area in southern Hillsborough for a long time are beginning to settle out, going into coveted construction or service jobs. Now a new group of workers is coming from different areas of Mexico, such as Chiapas and other southern states, filling the field jobs that the established Mexicans have left behind.

"But most of the families we work with are not settling out completely," Cannon says. "They are still traveling some in the summer to the Carolinas, New Jersey, and Michigan. Some even go as far north as Maine to pick blueberries."[3]

Angie Black

Since April 1999, the Presbyterian-sponsored Beth-El Mission in Wimauma is the home of a new, state-of-the-art facility for adult education headed by Angie Black. Black, who grew up in Ybor City, has worked with migrant and seasonal farmworkers since 1981. Though she makes her home in Pasco County, she is so dedicated to her work that she makes a fifty-five minute commute five days a week to facilitate the program. Beth-El sponsored a part-time adult education program for many years, but a new federal grant now funds a full-time program housed in four portable classrooms clustered around the property.

"We now have an entire portable classroom full of the latest computers, with top-of-the-line technology," says Black. She explains that with the help of the mission director, Evan Jorn, and Olga Ros, wife of the mission's Father Ramiro Ros, Beth-El pushed for a capitalization grant and was awarded over $100,000 to equip the adult classroom facility.

"The capitalization grant is for new and innovative programs. The money comes from the adult education grant, sponsored by the federal government under the Work Force Investment Grant," Black explains. "Hillsborough County gets $300,000 to provide job training and adult education through programs such as ours at vocational schools and at Hillsborough Community College. Adults are paid $3.35 per hour to go to school. They are also eligible for other financial aid, such as PELL grants and student loans.

"After they have graduated from high school or obtained a GED, we will pay the tuition to help them continue with college. We will also pay for tools for vocational training, if needed, so there are no out-of-pocket expenses. In fact, the student often winds up with money in their pocket, which helps them buy reliable transportation to get them to and from school and their job. This program has been around in Hillsborough County for over thirty years," says Black.

Black usually has thirty-five full-time students enrolled in the program. Other counties with large farmworker populations that have similar programs are Volusia, Dade, Collier, and Palm Beach counties. Black also says that she's noticed more interest in the education programs as more and more families settle out, leaving farmwork for new positions. With the economic boom of the 1990s, a glut of different jobs were available for people who normally worked the crops.

"Many farmworkers are settling out because they know it costs more to travel than to stay in one place," says Black. "The situation in Hillsborough is that there are now more jobs than workers. I've never seen this before, but I am actually getting calls from hotels and hospitals wanting workers. Before, they wouldn't want these former farmworkers, but now they are working in construction, in hotels, as waiters and landscapers, and in many other service jobs.

"Even the undocumented workers get jobs with fake IDs. I can spot one, but of course I won't turn them in [to immigration authorities] because they're only here to work. But I can tell a fake ID, even though they are very sophisticated these days. They come to your house, take a picture, and collect $30. Then they come back in about an hour, collect another $30, and give you a picture ID.

"I won't take undocumented workers with fake papers in my program because it is fraud, which means they can be deported and never allowed to enter the United States again. Our program is five hours a day, five days a week. They must verify they have worked in the fields in the last three years and show legal documentation to be eligible.

"It may sound like these people get a lot of benefits to continue their education, but they don't go to school just for the money," Black says. "I saw some of that in the 1980s, when the economy was bad, but they were mostly inner-city orange pickers. My students get preference at local vocational schools and community colleges because they are serious students.

"The most important aspect of our education program is sensitivity. Not everyone can teach these people because they are not sensitive to the culture. It takes time and perseverance. Many of these students are illiterate in their native language, so we have to teach them from A to Z. We divide them into three groups, with the basic group made up of those who do not speak any English at all, and use a California test to determine which class they belong with. But some students are not illiterate. Right now I have a student in the basic group who is a trained, highly educated psychologist, but he cannot speak a word of English."

Black also says most of her students are seasonal farmworkers who work locally harvesting tomatoes, cucumbers, peppers, cabbage and other vegetables grown in the area. Then they work the packinghouse season, spending the remaining part of the year on unemployment. As fewer families migrate, many are actually buying houses through Homes for Hillsborough, a program jointly administered by the Beth-El Mission and Hillsborough County.

Because her program is adult education, Black can enroll seasonal workers as well as migrants, and she also accepts young dropouts.

"There are some sixteen-year-olds who drop out of middle school because they have repeated the same grade three times. I can't blame them. It's usually a language barrier," Black explains.

Another important part of the education program is about domestic violence. Prior to Black's program, Beth-El incorporated the issue of domestic violence into the sewing and craft programs at the mission. Various volunteers from The Spring, Hillsborough County's domestic violence shelter, came to the classes and encouraged the women to discuss any domestic violence issues that may have resulted from the strain of poverty. Black says that since Hispanic culture is steeped in machismo, with males tending to rule the roost, there is often a lack of respect for women, who do not discuss these problems easily. Now domestic violence is part of the regular course for both men and women in the classroom.

Black credits the support from the Beth-El Mission for making the program so valuable: "The support they offer is phenomenal. They provide any assistance we need. When we set up the portable classrooms they looked bare, because there was nothing on the walls. I went with Olga, and she spent $300 to get things to decorate the walls. That made the opening wonderful."[4]

Margarita Romo

Farmworkers Self-Help in Dade City is typical of the programs in Central Florida that serve a well-established Mexican community where much of the farmworker population is more permanent. As in much of Florida today, the families travel less than they used to, staying longer in the winter months and venturing north for only a few months to harvest the summer crops. In these Central Florida counties, programs are de-

4.2. Tutors at Farmworkers Self-Help in Dade City teach workers to speak English. They also offer help for schoolchildren who have trouble with their homework.

signed to assist families, women, and children and to shore up the public school educational programs for school-age children.

The founder of Farmworkers Self-Help, Margarita Romo is a small, energetic woman with graying hair pulled back to an inconspicuous ponytail at the nape of her neck. She makes her home among the Dade City farmworkers, where she oversees this grassroots organization that assists the predominately Mexican workers who come to the area to pick oranges and strawberries. She is completely bilingual, easily interpreting the English words of a visiting county health worker to answer a Mexican woman's question. Romo says that of the many problems facing these workers, the most basic is the inadequate wage they earn because they are uneducated.

Born into a migrant family herself, Romo escaped the hard life when her father broke away from the grueling labor to become a gardener in Texas.

"That was hard work too," she says. "I worked with my father and so did my brothers."

When Romo married, she moved away from both the Mexican community and the farmworker community. But in 1971, she became involved again, and since then helping these workers has become her life. She has spent nearly thirty years working with Florida's migrant laborers and traveling the state as an advocate. She also served on the governor's council for helping farmworkers in 1998, but the council was disbanded when Governor Jeb Bush took office in 1999.

Romo well understands the value of education for these children, and she has set up a program in her small Dade City center to help keep farmworker children in school. The original building had a meeting room and a small adjoining room with two old computers donated to the students. There were also pictures of Christ, because Sunday services were held in the busy building. Now there are several buildings, with a clinic and one solely devoted to education, as well as the mission.

"We're trying to get more done in the school system but it's hard," Romo says. "We have an agreement with the Department of Juvenile Justice. We wrote a proposal [for what] we call the 'Dream Team.' We

4.3. Margarita Romo of Farmworkers Self-Help in Dade City teaches social justice to members of the "Teen Dream Team," a group of teenagers from the farmworker community who want to improve their education. They meet once a week in the multipurpose building housing the grassroots organization founded by Romo.

4.4. Blanca Padron has taken all of her six children to fields to work. She and her husband, Hector, travel to Texas each summer when the children are out of school, because there the children receive hourly pay at age thirteen. In Florida, they earn by the piece.

[Farmworkers Self-Help] want to be a liaison between the kids and the system, and we are. Two of us work on that project. When there's a suspension in school, we go and try to get it lessened or we teach the suspended child in our center.

"We jump in and say, 'what good is this gonna do the kid?' I just had a little girl sixteen who did five days with us just so she could go back to school. We did tutoring while she was with us. We taught her to answer the phones around the office. She was bright.

"We hold a meeting of the Dream Team every Wednesday night so the students can discuss their progress and any problems at school or at home. The Dream Team is a group of teens in the farmworker community who study hard and want to get ahead. We encourage them to stay in school and get an education. The teens in the program usually succeed.

"There are about twenty kids in the program. They walk here and they come voluntarily. We're taking them to Tallahassee on spring break this year so they can see how the legislature works."

Romo also visits the schools and the parents. When there is family violence, Farmworkers Self-Help offers counseling for the family.

"Kids act out in school because they're frustrated at home," says Romo. "Farmworker kids get tired of coming to the school session late in the fall and leaving early in the spring because their families have to move to different fields around the country. The oranges are done now, and they have to move up to South Carolina, to new schools and new teachers.

"The kids often don't speak good English, so many of them eventually just give up. We have volunteer tutors at the help center, and little by little they bring the children in for assistance. We give one or two $500 grants every June. We find one kid who wants to go to college and they go."[5]

Migrant children also gain a lot of worldly knowledge before they enter school. Their experiences are broad as they travel the country, seeing many places and learning how to fend for themselves at a very early age. What they learn at school is often quite at odds with what they learn at home, where cultural values and habits may be very different.[6] Though there may be problems for many of these children, there are also victories that make the struggle worthwhile.

Ninfa Griffin

At thirty-four, Ninfa Griffin is a real success story, the kind of success that makes advocates and farmworkers alike proud. Griffin teaches second- and third-grade students in the English for Speakers of Other Languages (ESOL) program at Dover Elementary in the rural community of Dover in southern Hillsborough County. For seven years she has taught migrant children, encouraging them to learn and trying to elevate their self-esteem. In 2000, Griffin represented Florida when she received the Distinguished Graduate of Title I for 2000 award, in San Antonio, Texas. The award is given to a successful person who benefitted from the Title I program. Griffin is keenly aware of her students' experiences, because she was born into a farmworker family in San Juan, Texas.

"I worked in the fields with my family until the tenth grade, and now I'm an elementary school teacher teaching migrants," she explains. "I moved to six different schools every year. We came here in the late seventies to pick oranges, and it became a base. Every year we traveled to Michigan, Indiana, Ohio, Virginia, and Tennessee. We worked picking strawberries, cherries, tomatoes, and asparagus. To me, asparagus was the best. I really liked it because it grew in very sandy places, and I thought it was fun because it was like the beach. Sandy. I didn't have the water, but I had the sand.

"I started working in the fields at age five, mostly because there was no day care. My parents would pick apples, and they would put us in the apple boxes to pick the leaves off the apples. Simple stuff. When I was older I picked everything—tomatoes, peppers, cucumbers, all the vegetables."

Today Griffin shows her students pictures of herself picking vegetables, and she tells them that they too can become teachers if they just focus.

4.5. Children become used to farmwork at an early age because they sometimes help in the fields. Here a woman and her daughter pick strawberries side by side. Though children are not allowed to work during school hours, they can work on Saturdays and after school.

"Most of the families are very receptive to educating their children. They want to have a better life and get away from farm labor. Teachers are highly esteemed in Mexican culture. Even though the parents are illiterate, they want their kids to succeed. I tell them they really need to focus.

"Most of my students are monolingual, speaking only Spanish. This year I have eight students from Oxaca who have just come here. They speak only the Indian dialect, which I don't speak at all. I really am dependent on several bilingual students who speak both the Indian dialect and Spanish. They can tell these students what's going on and explain what they should do in the class. Last year I had a girl from Guatemala who spoke the Indian dialect called *Mann*.

"Moving is the worst problem for these kids. Five of my students have left since Christmas. Sometimes they don't even know they are moving until it happens. I try to put a package together for them to take with them—something with some of the stories they have written, some pencils and paper so they can continue drawing or writing.

"As a Title I school, we are only allowed to have twenty-one students in a class. Eight kids stay almost all year, but twelve are only here for about three months, maybe five months if you stretch it. It makes my job hard because I want to teach them practical things, but they must move with their families. The majority of the children return the following year."

Griffin also visits the homes of her students, usually trailers, to talk to parents and other family members and discuss any problems the child is having.

"There are so many people," she exclaims. "Sometimes there are three families in one home. They even pull out mattresses and sleep in the kitchen. In my class I sometimes cook with the children, and they don't even know what the oven is for. They think it's storage for pots and pans, because most of the ranges in their houses don't work. A few years ago I had one student whose mother had a working oven. She was so proud; she baked cakes for any occasion.

"The children are very spatial. They really like art. And my classroom is always clean because they are taught to be clean. They clean the house at home while their mother is working. They may have meager means, but they are very clean. The camp may be nasty outside, but inside it's clean.

"Some of the families who have just come from Oxaca are not so clean. They may not be used to running water, so we encourage them to bathe and be clean. One little boy kept coming to school with very dirty fingernails, and he didn't look clean. We went to his house and explained to the father that he must bathe if he's coming to school. Since then he's been cleaner. Some of the older girls have to be reminded to wear deodorant.

"It's often like Mexico in the camps. In Mexico, the government doesn't provide facilities for garbage. You must pay a man with a donkey to come and haul it away. We try to teach the parents not to throw trash in the yard, like they do in Mexico."

Another problem Griffin notes is the toilet training of some of the children: "In the camps where they live, the farmers who rent the housing tell the farmworkers not to put paper in the toilets. In their homes, they have a bucket next to the toilet where they are supposed to put the paper. When they come to school there is no bucket or trash container in the toilet stalls, so they sometimes put the toilet paper in the corner. We have to teach them to flush the toilet paper down the toilet.

"We also work on their self-esteem and teach them to become productive citizens. They are very receptive. We also teach them rules. In Mexico, if someone sees a fishing hole, they just drop a line in. Here we teach them that they might be trespassing if they fish in a pond. We tell them they must ask."

Griffin says that there is little child abuse in the farmworker communities because they love their children. She recalled one case a couple of years ago they investigated.

"We had a little girl who came to school with a bruise on her back. We went to the home and found out it was her older sister who beat her with a belt because she didn't do the housework. They all have their jobs to do. They're such little mothers. We talked to the family and the sister and told them it was wrong to beat the little girl with a belt."[7]

There are examples of success stories in migrant education, but each one is a difficult victory. If the family is traveling, the student may attend as many as six or more schools a year. Friends are temporary, and the trauma of being very different is real. These children rarely go to movies or museums, and, if they are part of a migrant family, they seldom participate in school sports. Educators often want to categorize, but these children have their own category, which is difficult to judge by middle-class America's very different standards.[8]

CHAPTER 5

Health and Safety

The walls of the modest office of Farmworkers Self-Help in Dade City display several posters in Spanish. On one, a photo of Rosalyn Carter smiles down, urging parents to vaccinate their children. Another cautions against drugs, and still another against AIDS; both are accompanied by cartoonlike drawings to illustrate the message. These posters are typical in most advocate centers for farm laborers. Some are in Creole as well as Spanish, so that Haitian workers can understand them. There are educational pamphlets, too, advising caution when handling pesticides. All of this information is geared to protect those who harvest the crops for America's tables.

In addition to the long hours, harsh work, and hazardous conditions that threaten their health, when farmworkers leave the fields they often go home to unsanitary conditions. Nutrition may be poor, leaving them at risk for deficiency diseases, and farmworkers have higher rates of parasites and certain infections. The life expectancy of migrant and seasonal farm laborers is nearly thirty years lower than that of the average American, while the infant mortality rate is double the national rate.[1]

Many health-care professionals who treat farmworkers compare their living conditions to those in developing nations. In 1981, of 400 Florida labor camps that were investigated, 40 percent did not conform to minimum health and safety standards. Especially in trailers, sewage systems

5.1. No one was at home during the day at this labor camp in Pierson, but several scruffy dogs kept guard amid clutter and garbage. The houses were obscured from the road by a thick band of scrubby trees. Alfredo Baheña, an advocate with the Farmworker Association of Florida, said inspectors seldom visit such areas, though they are supposed to inspect housing once a year.

are often inadequate, leaking, backing up, and sometimes draining improperly. Wiring may be unsafe, exposed, or poorly grounded, creating a fire hazard. Many dwellings lack adequate heat for cold weather.[2]

Sanitation in the fields is another major issue, but in recent years conditions are better. In 1970, agricultural employers gained exemption from the worksite protection act imposed in other industries. Farmworker advocates pressed the Occupational Safety and Health Administration (OSHA) to establish a federal standard, but it was not until 1985 that Secretary of Labor William Brock gave states eighteen months to establish federal standards. One of the things instituted was proper field sanitation, which means available portable toilets, fresh water and disposable cups, and facilities for washing hands to remove pesticide residues before eating. Lack of clean water contributes to heat stroke from dehydration, and tuberculosis or trench mouth can result from sharing drinking vessels. Hepatitis, amoebic dysentery, typhoid fever, and bladder infections may spread rapidly through close contact and unsanitary working conditions. Though some of these diseases may not be life threatening, they can become chronic and uncomfortable without diagnosis and care.[3]

Tom Himelick

Tom Himelick, the associate director of the Emory University Physicians Assistant Program, goes to the fields of Georgia and South Carolina each summer to train graduates as physicians' assistants. Himelick and his students visit farm sites to assess the quality and availability of health care provided for the migrants moving north from Florida. They make it their mission to bring primary health care to rural and underserved areas in collaboration with the Southwest Georgia Area Health Education Center. Himelick and his crew go onsite to the fields, but they offer only primary care, working with limited equipment and doing most surveys at the side of the field and outside the packing sheds. He describes the major health complaints of farmworkers.

"We see many musculoskeletal problems, such as back pain, knee, and muscle pain, from stooping and lifting. There are also eye complaints,

5.2. Portable toilets located near the edge of the fields comply with health regulations. Crew bosses move them from field to field with the workers, but sometimes the crew bosses are slow and the crew is finished picking before the crew boss arrives with the portable toilet. Because the toilets are not always clean, they may contribute to bladder infections in women.

5.3. These three toilets sit at field's edge, but some fields are so large that it may take a quarter-mile walk to reach them. The toilets can become like ovens as they bake all day in the hot Florida sun.

mostly from pesticides and dust. Workers also suffer headaches, possibly from heat exposure and dehydration. Skin problems, such as contact dermatitis from chemical or plant sources, is often due to improper clothing and shoes in bad shape. Shoes that don't keep their feet dry encourage fungal infections," he says.

"There were also some gastrointestinal complaints," Himelick continues. "We're not set up to look for parasites. Other reports have indicated a high incidence of parasites among migrants due to sanitation, but we have no lab equipment with us in the field."[4]

Women's Health Concerns

Women have special health concerns. When asked what troubled them most, women cited such issues as pesticides, AIDS, and a higher than average incidence of birth defects among children of workers. Continual exposure to pesticides can have devastating effects on women who are pregnant or in their childbearing years. Inadequate toilet facilities in the field leave women more susceptible to urinary tract infections, which, when contracted during pregnancy, have been linked to increased risk of miscarriage, premature labor, and neonatal death. Other studies cited an incidence of cervical cancer among women in their twenties that is much higher than the national average and a high rate of leukemia among children of farm laborers that most health workers attribute to pesticide exposure. Hindered by lack of medical insurance, information, and access, few women farmworkers are able to obtain regular health care for themselves and their families. They may suffer needlessly, and some even die from preventable, treatable conditions.[5]

Current regulations require growers to provide portable toilets that move with the workers from field to field. They must be near the edge of the field, and they must be cleaned each day. Faucets with running water are available on water trucks that allow workers to wash their hands. This is a great improvement over conditions of twenty years ago, when workers just squatted in the fields. But in spite of the requirements, there are still some fields where pickers stoop for hours gathering the harvest with

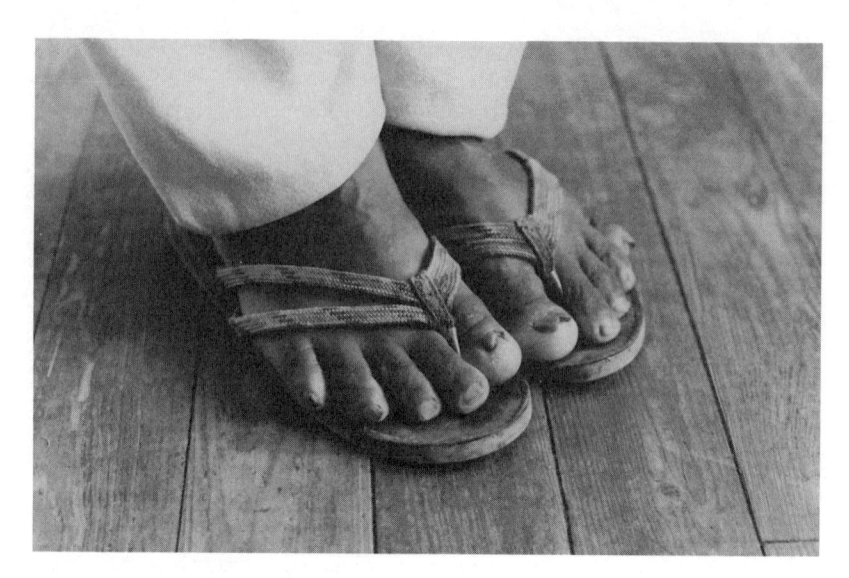

5.4. Fungal infection of the feet, caused by improper footwear and constant dampness in the fields, can lead to uncomfortable foot problems. This is one of the most persistent and widespread problems that health workers see among farmworkers.

no portable toilet in sight. Even today, some women still wear skirts over their work pants so they can relieve themselves in the open fields when the portable toilet is an acre or more away at the far edge of the field.[6]

Disease

AIDS and other sexually transmitted diseases are higher than average among farmworkers, as are musculoskeletal problems, fungal infections, and urinary tract infections. Drug and alcohol problems are mostly confined to the younger workers and those who travel alone. Teen-age children may be more susceptible to substance abuse, as in any underprivileged group. Recently, tuberculosis has increased among farmworkers, which spreads easily through their close living quarters. Other diseases listed by the Center for Disease Control (CDC) as prevalent in these communities are diabetes, hypertension, and several communicable diseases uncommon among the general population. Yet despite these problems, the Office of Migrant Health in the Department of Health and Human Services estimates that only 12 percent of the farm workforce in the United States receives help from federally supported health centers. However, privately funded clinics sometimes have more outreach programs and are less strict about demanding legal paperwork that many workers may not have.[7]

At the CDC in Atlanta, disease prevention is the prime concern, especially the prevention of AIDS and tuberculosis among farmworker populations. According to the CDC, farmworkers are approximately six times more likely than the general population to develop tuberculosis. The disease may be asymptomatic, so testing is highly important. Currently the CDC is chiefly concerned with drug-resistant tuberculosis, because it requires different treatment than the less complicated pulmonary tuberculosis, which responds easily to such drugs as streptomycin. Higher rates of the resistant strain occur in the ethnic and social groups that comprise much of the migrant workforce. The CDC recommends careful monitoring of patients infected with the resistant strain and, ideally, placing them on "directly observed therapy given by a well-trained out-

5.5. Constant stooping and the lifting of heavy buckets often leads to arthritis and other musculoskeletal problems among farmworkers. Here a woman bends down as she plants cucumber seedlings in dirt mounds covered with plastic mulch. Each row may be 200 to 300 feet long. The bandana under her hat helps keep her cool.

reach worker from the same cultural/language background as the patient."[8]

The CDC's Advisory Council for the Elimination of Tuberculosis conducted the first population-based study of tuberculosis in a random sampling of North Carolina farmworkers. The study found active tuberculosis in 0.47 percent of the Hispanics and in 3.5 percent of African Americans. Because the bacillus spreads through the air when an infected person coughs, the crowded living conditions and close quarters in vehicles create ideal circumstances for transmitting tuberculosis in migrant communities. Sometimes the disease remains undetected for years, and persons with a compromised immune system, such as those with HIV infection, are more receptive to tuberculosis.[9]

Because of the difficulties inherent in tracking, treating, and isolating people with tuberculosis in the farmworker community, the CDC recommends that migrant health-care providers actively pursue these traveling harvesters. Most rural clinics offer consultation and testing, provide free hospitalization to infected farmworkers and their families, isolate them if they are infectious, monitor them for the course of the treatment, and teach them preventive measures. If farmworkers leave to harvest in another area while undergoing treatment, health-care workers contact tuberculosis control officers in other states to keep tabs on these individuals. However, the CDC has a strict privacy policy and only encourages sharing this information on a need-to-know basis. Though the occurrence of tuberculosis in farmworker families impedes both their ability to work and their social progress, it is often difficult to convince them to stay in touch with health officers. Many infected workers slip through the cracks, remaining a danger to themselves and others.[10]

In Belle Glade, the high incidence of AIDS among Haitian workers once led to fears and the circulation of the unfounded rumor that Belle Glade was the "AIDS capital of the world." But today, due to education and treatment, the disease is in decline in the community. AIDS deaths in Palm Beach County, where Belle Glade is located, have dropped dramatically since 1996, when the county was second only to urban Dade County in the number of AIDS-attributed deaths. In 1998, AIDS deaths fell from the 1996 total of 306 to a low of 135. The state offers anonymous testing for HIV, and clients receive information on prevention, the benefits of early treatment, and referrals to care, along with other needed services. In addition to confidentiality, the Florida Department of Health offers several help programs for people living with AIDS and HIV. The Department of Housing and Urban Development (HUD) funds Housing Opportunities for Persons with AIDS (HOPWA) through a grant that provides states and certain metropolitan areas with resources for meeting the housing needs of persons with HIV disease and AIDS.[11]

The Florida Department of Health contracts with agency organizations at the local level to administer these programs. These emergency services provide transitional housing, rent, utility payments, and other necessary expenses for people with the disease who are unable to work. There is also drug assistance through Florida's AIDS Drug Assistance Program, which provides medications to those with HIV or AIDS whose income falls below the poverty level. These programs are funded through Ryan White Title II grants. However, workers lacking the proper documentation do not qualify for these government-sponsored programs and must fend for themselves.[12]

Florida also sponsors the Vaccines for Children program, which allows eligible children to receive free immunization for childhood diseases. The program is funded by the federal government and is available to children on Medicare, Native American children, and children without insurance. Any pediatrician or private health facility that serves children can obtain the vaccines. Outreach workers try to reach even the most remote farmworker families in order to get the children vaccinated, because farmworkers' living and working conditions make them especially vulnerable to these illnesses.[13]

Caridad Ascencio

At the Caridad Health Care Clinic in Palm Beach County, farmworkers receive both medical and dental care. The clinic is named for Caridad

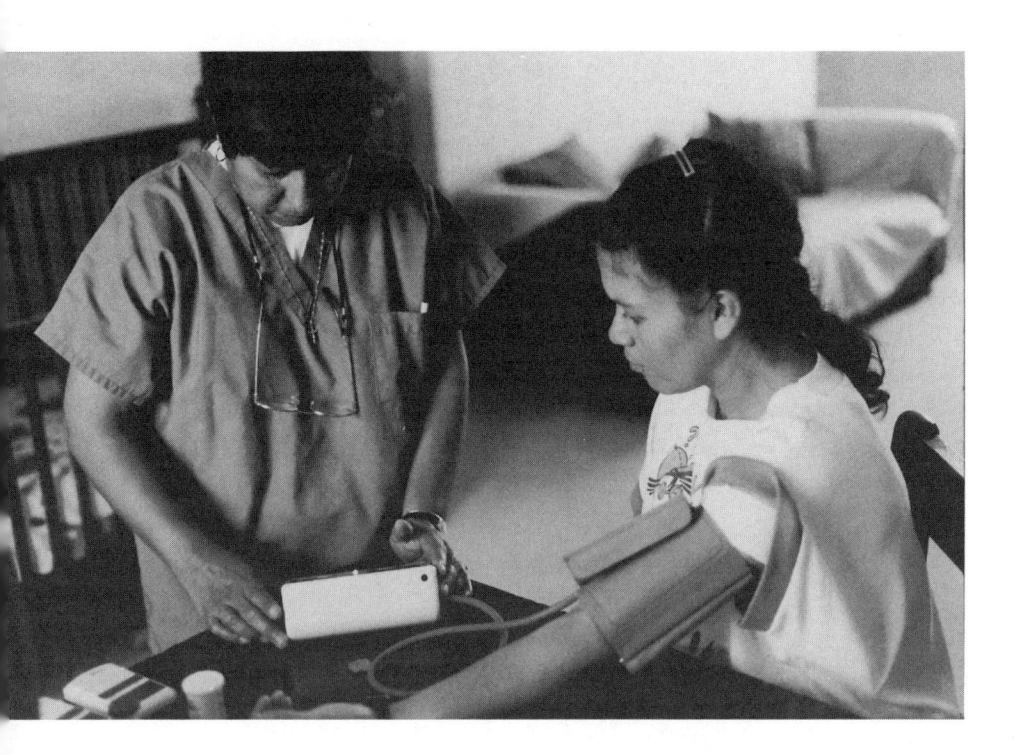

5.6. An HIV outreach team member takes this woman's blood pressure. Transportation to clinics and the limited hours of operation are problems for most farmworkers, and this may be the only contact they have with a health-care provider. In addition, many clinics may have no one who understands the workers, especially if they speak an indigenous dialect, as do many of the Guatemalan workers. Some outreach workers have more contact with farmworkers than do the county health agencies. Since farmworkers toil long hours on weekdays, many doctors are willing to work in clinics on weekends.

Ascencio, who founded it in 1992 to help children of farmworkers obtain health examinations and vaccinations before entering school. Ascencio, a Cuban native who came to this country in 1961 as a political refugee, was a community health worker for more than twenty years before she decided to start her own clinic.

Originally the facility was in a double-wide trailer, but the growing demand for the clinic's services led in 1997 to a new 7,300-square-foot facility and to its services expanding to include dental care. In 1999, the clinic, which now serves both children and adults, treated eleven thousand patients. Today the majority of the clinic's patients are Mexican Hispanics, but beginning in the 1980s, a large influx of Central American refugees arrived to join them as farm laborers.

"Lately we see more and more Guatemalans, El Salvadorans, and Nicaraguans," says Ascencio. "They have left because of the political problems in their countries. We also have Haitians, Puerto Ricans and Jamaicans.

"Most of the injuries the clinic treats are farm-related, such as eye problems from chemicals. Recently we treated two men with eye injuries from chemicals. One man we could treat, but the other man was blinded. We sent him to the hospital, but they could not save his sight. We think it was from fertilizer, but we are not sure. We also see a few cases of pesticide poisoning, but not too many."

Ascencio agrees with other health professionals that most cases of pesticide poisoning go unreported because the symptoms are so nebulous. Symptoms resulting from chemical exposure—headaches and muscle aches, rashes, nausea—could be anything, and a farmworker unfamiliar with the dangers of pesticides may not be aware that health problems could be tied to these toxins.

"AIDS is also a concern, especially among the large population of single men who live in the three labor camps that surround the clinic. Recently we got a grant to stay open one night a week so we can check the men's camps for AIDS," says Ascencio. "I would like to be open three nights a week, but we can't afford that. We want to check all the men in these camps around here for AIDS, because we can see them at night. We

5.7. Trash and garbage are common sights around farmworker housing because the camps are often far off the main roads. Here the owner of the trailers has hired a waste removal company, but garbage piles up before the garbage truck comes. Transient workers seldom come home except to get some sleep before they rise at dawn the next day. (Photo by Nano Riley)

also see lots of diabetes among adults," says Ascencio, "and lots of high blood pressure."

Most of the AIDS cases seen at the Caridad Health Care Clinic are among adults; Ascencio has not seen cases of children born with the AIDS virus, though she cannot be certain. The clinic tries to test all of the children for tuberculosis, especially those who have been exposed, and to give them preventive treatment.[14]

Children

Discussions about farmworkers inevitably turn to issues of child labor. Certainly there are laws strictly regulating the ages at which children or teens may work. No matter how stringent the regulations, though, there are always youngsters who find ways to work. Some of the young, single Mexican men who enter the United States to earn money bring falsified identification stating they are sixteen or older. They memorize birth dates so they can answer quickly, and some bosses never ask if the young man looks old enough to work. Mustaches, popular with Hispanic men, may add a few years to the appearance of an underage teen who wants to add to the family income by working.

Farmwork is an exception to the rigid laws against child labor in the United States. A section of the Fair Labor Standards Act, the law regulating child labor, prevents those under sixteen from performing what the Department of Labor determines is hazardous work. However, children aged fourteen to sixteen may work in certain instances, particularly on a family-owned farm. Some of the occupations considered too hazardous for children are operating most heavy machinery, such as tractors over twenty horsepower, or assisting in the operation of mechanical corn pickers, grain combines, and hay balers. Children are also restricted from operating any earthmoving equipment, forklifts, or power-driven saws. The law also mentions that those under sixteen may not work in a fruit or grain storage facility "designed to retain an oxygen deficient or toxic atmosphere."[15]

Child labor is certainly a major concern of farmworker advocates, who have long recognized the problem of children suffering injuries while living or working on or near farms. Recent statistics indicate that about 100 youths under twenty die on farms each year as the result of accidents, and more than 100,000 related injuries occur to those under twenty. The media have widely publicized the issue of children working on farms, and many advocate groups have protested it, but no national coordinated effort to address the problem existed until the 1990s.[16]

In April 1992, a symposium on childhood agricultural injury prevention took place in Marshfield, Wisconsin. Sponsored by the National Farm Medicine Center, the symposium was intended to develop research, education, policy, and other interventions to reduce agricultural injuries among children. A core of forty-two individuals present at that meeting formed the National Committee for Childhood Agricultural Injury Prevention (NCCAIP) to review relevant information from previous injury reports and data. In April 1996, NCCAIP published a National Action Plan promoting the health and safety of children exposed to agricultural hazards. Now NCCAIP is conducting surveillance of all childhood injuries across the country in order to reduce injuries to children working in and around agriculture. The group also identified jobs performed by children on farms that create stress on the musculoskeletal system and studied new ergonomic innovations designed to lessen potential injuries. Tractors operated by youngsters are another hazard, so the group produced a CD-ROM titled *Kayle's Difficult Decision,* which offers a realistic scenario about an inexperienced teenager who learns to be responsible when operating a tractor.[17]

Contacting young people at risk for farm injuries is difficult. In interviews conducted in Homestead, Florida, health officers found that few adults were willing to admit that children or adolescents in their households were involved in farmwork. That made the original goal of using community-based surveys unrealistic. Because the adults did not admit youngsters were working, it was impossible to conduct interviews with the children themselves, a tool that health officers needed to implement

5.8. A mother and daughter pick tomatoes side by side. Because it is not during school hours, the child is allowed to help. Although she probably receives no official pay, she adds to her mother's harvest.

proper safety precautions. Further complicating the problem is a Department of Labor study that indicates that most adolescents doing migrant and seasonal farmwork in this country are the single, undocumented Mexicans, some of whom may be as young as fourteen. They remain undetected because they never enter school or participate in community activities. These young men remain virtually unreachable, and the crew bosses want to keep them in the shadows.[18]

Employers can apply for waivers, allowing them to use ten- and eleven-year-olds for hand-harvested produce if they can show the crop has a short harvest season, and that not using these children would cause great economic harm to the industry. Employers must also prove the work will not affect the children's health, and that the youthful workers will not be exposed to pesticides and other chemicals causing adverse health effects. Restrictions prevent the children from working during school hours, but youngsters of twelve and thirteen can work outside of school hours in nonhazardous jobs if the work takes place on the same farm where a parent works.[19]

Florida law states that children are allowed to work in the fields from the time they are fourteen, but only on weekends or after school. They cannot perform hazardous work, such as operating farm machines or handling pesticides. At sixteen, they may work full time, but until they are eighteen, there are still restrictions on certain work considered dangerous.[20]

Margarita Romo often sees children working in Dade City. "We have a children's church here at the mission, and during the orange season, many of the kids weren't there," Romo acknowledges. "Twelve-year-olds have to get out of school to help their parents. One sixteen-year-old I know works in the fields instead of going to school. If you see these kids are absent more than they're in school, you know they're going to fail. They're not supposed to work in the fields but they do," she says. "Little kids too young for school hang with their mothers. Often there's no day care."

Because day care may be unavailable or unaffordable, mothers take the little ones with them into the fields. Even if they are not working, the children will play along the side and risk exposure to chemicals and hazardous conditions. Even with laws in place, children often work for a couple of weeks here and there at peak harvest season, and everyone just looks the other way.[21]

Farmworker children sometimes play around heavy machinery or help out in the packinghouses. There are frequently news stories about children injuring themselves playing near tractors or other farm machinery. In 1997, the Maternal and Child Health Bureau, in conjunction with the National Institute for Occupational Safety and Health (NIOSH), founded the National Children's Center for Rural and Agricultural Health and Safety (NCCRAHS). The group studies the health and safety of children who may do agricultural work in rural areas in order to identify risks and to educate children and parents in ways to handle these hazards.[22]

Safety in the Fields

The most comprehensive study of farm injury morbidity and mortality to date is the National Safety Council's 1988 survey of 127,169 farm family members, which included 57,301 full- and part-time employees on 37,293 farms in thirty-one states. The data covered more than 5,753 injuries, ranging from minor problems to crippling and fatal accidents. The survey grouped farmers and farmworkers together, so there is no distinction between the two groups.[23]

The highest work-related injury rates were in the age group of five to twenty-four years. This group had a combined overall rate of 25.8 work-related injuries per million work hours. Work-related fatalities for those under fifteen were over 7 percent, quite high considering that this age group performed only 4 percent of the combined farm working hours nationwide. Fatalities were also higher for workers over age sixty-four, who perform 11.1 percent of the work and contribute 5.5 percent of the total work hours. Agricultural machinery was the single leading source of occupational injury, causing 17.6 percent of total injuries, followed by animal-related injuries at 16.9 percent. Other studies of occupational in-

5.9. Sharp knives are essential for cutting cabbage heads and trimming tomato vines. Here a man sharpens his knife as he prepares to cut the strings from the spent plants in this tomato field so it can be cleaned up for replanting.

5.10. The workers' enemies are rains and freezes. This woman protects herself from the rain with a makeshift poncho fashioned from a plastic garbage bag. Hard rain can make it impossible to pick, especially in the summer when there is often lightning. In 1999, two workers died when a bolt struck them as they ducked under a truck parked in the fields to seek shelter from a rainstorm. Freezes may ruin the vegetables. These unpredictable natural occurrences threaten an already shaky income.

juries and mortality in farmers and their dependents show similar results, but, again, there was no distinction made between farmers who own their farms and the workers they hired.[24]

The most frequent causes of farm accidents for those working with field crops are falling stacks of crates, overturning gondolas, and accidents associated with farm machinery—forklifts and tractors. Other common causes of injury are tree accidents caused by breaking tree limbs or by workers falling down from ladders with bags full of fruit, resulting in fractures, sprains, contusions, puncture wounds, and lacerations. Of 237 migrant farmworkers studied in North Carolina, 24 reported an occupational injury during the previous three years. Broken bones, sprains, and cuts accounted for 80 percent of the injuries. Vehicles or machinery caused 21 percent of the injuries that caused time lost from work. The relatively small number of known injuries does not give a true picture of farmworker safety, because many accidents go unreported. The incidence of injury (8.4 percent for a three-year period) is, however, probably an underestimate.[25]

For several reasons, data on farmworkers' injuries are not as readily available as that for workers in other industries. First, for farms with fewer than eleven employees, there are no legal requirements to report injuries. Farms with eleven or more employees must follow the regulations of the Occupational Safety and Health Administration (OSHA). Workers' compensation data for agriculture are not consistent or even widely available because of the many exemptions and loopholes in state laws. Second, farms are often in distant rural areas, so it can be expensive to collect data from some locations. And again, there is the ever-present problem of illegal workers, who may not even seek treatment unless the injury is severe.[26]

Long-time farmworkers learn farm safety. They often move up to become crew bosses or equipment operators, jobs where experience counts. But there is a constant supply of immigrant laborers who speak little English and find themselves in tedious, disagreeable, and often unsafe jobs that others refuse.

CHAPTER 6

Pesticides

At the edge of a fallow field in southern Hillsborough County near the tiny hamlet of Parrish, empty pesticide containers lie in a heap, decaying in the sun. The names of the chemicals are now illegible, but the skull and crossbones symbol on the fading labels is a dead giveaway for what they once contained. Some are plastic jugs; some are cardboard boxes.

Although farmworkers know pesticides can be harmful, many are unaware of the long-term effects and how they may affect their children, especially the unborn. The symptoms of exposure to these poisons may be similar to those caused by a flu or cold; sometimes it is a persistent rash, a bit of numbness, or headaches. Most do not even bother reporting such nebulous symptoms. With the need to put food on the table, many workers pay scant attention to aches and pains that might cause others to miss a day of work.

Florida passed the "Right to Know" bill in 1994, which allowed workers to know what they were handling. Prior to the passage of this bill, no law required growers to notify workers of the potential dangers of any pesticides they handled or even what the pesticide was. Though the law states that pesticide applicators must take a course in pesticide safety and receive a certificate allowing them to use the commercial-strength chemicals, the poisons may still be applied by anyone under the certified person's supervision. This means that anyone can apply pesticides, even if

6.1. A huge pile of garbage, consisting mostly of discarded pesticide containers, litters the side of a field in Central Florida. An electric pole with a light aids workers in the early morning hours before sunrise. The round tank to the left of the pole is for storing pesticides. There are rules against dumping pesticides, but the residue that may leak from the containers does not seem to bother anyone.

6.2. Lack of child care often means that children must accompany their parents to the fields. Here several older siblings watch the younger ones at a field's edge. Farmers may turn a blind eye, because if they prohibited children near the fields, crops might not be harvested in time. While in the fields, the youngsters can be exposed to excessive levels of pesticide, but often parents have little choice but to bring them.

the worker doing the actual application cannot read the instructions that are printed in English.[1]

Because pesticides are easy to mishandle, there are frequent newspaper accounts of poisoning among unsuspecting farmworkers. Sometimes these poisonings lead to lawsuits against the largest chemical manufacturers, but it can be tough business fighting such giants as DuPont, Monsanto, and Dow. The difficulty of pressing these suits was illustrated a few years ago by the case of Juan and Donna Castillo, who sued DuPont and Pine Island Farms when their son was born without eyes. Donna Castillo claimed that in November 1989 she was soaked with the pesticide Benlate by a tractor spraying tomato fields belonging to Pine Island Farms as she walked near her South Florida home. At that time she was seven months pregnant with her son, John.

In 1996, a state court jury in Miami deliberated the negligence case brought against DuPont alleging that the Castillos' son's birth deformity was caused by the fungicide Benlate 50 DF, which was commonly in use in vegetable fields. Attorneys for DuPont and Pine Island Farms argued that Benlate was not in use at the time. DuPont claimed that its product, accused of causing devastating damage to some farms by killing plants, causes no damage to humans. The case was the first claim involving the boy's birth defect, microphthalmia, or tiny eyes. (In John's case, there were only dents where his eyes should have been.) In June 1996, a Miami jury awarded $4 million in damages to John, with nothing going to his parents.[2]

Since 1991, the year DuPont recalled Benlate, roughly 500 cases, many of them in Florida, have been tried in state and federal courts around the country. There are also twenty-five cases pending in Scotland, where farming communities blame the chemical company for children born without eyes. Most cases involved farmers and nursery operators who claimed that the DuPont product wiped out their vegetables and ornamental plants. When DuPont's own researchers decided their product was not to blame, the company decided to fight these claims.[3]

The chemical company has paid out more than $1 million in damages. DuPont was a codefendant with Pine Island Farms in the Castillo case.

Pesticides   129

The tractor, the Castillos' lawyers said, got stuck in neutral and sprayed the mother as well as her young daughter. The farm was accused of spraying in excessive wind. The EPA has standard regulations that prevent spraying if the wind is blowing more than ten miles per hour. In a twenty-nine page opinion, the Third District Court of Appeals said key expert testimony linking Benlate to birth defects should not have been admitted. Benlate was banned from use on American crops, but other pesticides still are causing problems.[4]

Methyl bromide is now a major political issue in Florida and California, where farmers use the chemical to grow the most popular fruits and vegetables brought to our tables. In Florida, growers use methyl bromide in growing tomatoes, strawberries, and bell peppers. Methyl bromide is used as a soil fumigant before planting to eliminate nematodes and sterilize the soil. Because it is completely odorless, it is mixed with tear gas to prevent exposure in farmworkers. In California, methyl bromide poisoning is the fourth leading cause of injury among the farmworkers who use it. In 1998, the California Environmental Protection Agency blamed fifteen workers' deaths on the poison and reported that another 216 suffered illness from exposure to methyl bromide. In 1995, the Environmental Protection Agency (EPA) halted the testing of methyl bromide on beagles when they found the dogs were suffering severe neurological damage that caused them to bang their heads into their cages. In spite of the neurological damage it can cause, big berry growers are so dependent on methyl bromide as a soil fumigant that they refuse to stop using the chemical.[5]

Methyl bromide is not just harmful to humans and animals, it is also destructive to the Earth's ozone layer. Recent studies in California show that methyl bromide hangs heavy in smog, and that it often drifts far from the initial point of application. The EPA issued a directive to stop using methyl bromide by the year 2001, but many growers requested extensions.[6]

In 1998, the environmental group Friends of the Earth worked with the Farmworker Association of Florida, Farmworkers Self-Help, Florida Consumer Action Network, and the Legal Environmental Assistance

6.3. These boards, required by law, stand at the entrance to most fields. Notices let workers know when fields were last sprayed, when workers can reenter fields, and what to do if they think they have been exposed. The bulletin boards also post the name of the farm, the basic regulations governing children working in the fields, and other items most workers may need to know.

6.4. Methyl bromide, a powerful fumigant sprayed heavily on strawberry and tomato fields, is being phased out by the EPA, but growers may use it for several more years. The odorless poison must be mixed with tear gas so that workers will know if they are being exposed.

Foundation to issue a report on methyl bromide use in Florida. The report, titled *Reaping Havoc: The True Cost of Using Methyl Bromide on Florida's Tomatoes*, documents the effects of the toxic gas on farmworkers and their families, as well as the potential damage to communities located near treated fields and the chemical's destruction of the Earth's ozone layer.[7]

More than 160 nations, including the United States, agreed to phase out 50 percent of methyl bromide use by the beginning of 2001. The agreement, known as the Montreal Protocol, was reached in 1987 and has been revised several times, the most recent in 1999. It encountered strong opposition in Florida from the Florida Fruit and Vegetable Association, the Florida Farm Bureau, and the Crop Protection Coalition and nationwide from the California Strawberry Commission and the Great Lakes Chemical Corporation, a major manufacturer of methyl bromide. Agribusiness argues that there are no alternatives to methyl bromide. The proposed schedule of the EPA gives growers until 2005 to complete the phase-out, but growers continue to fight the ban.[8]

While the fumigant does not last long on the tomatoes and berries that grow in the treated earth, methyl bromide does kill anything in the soil, including beneficial bacteria and the valuable earthworm. The Florida task force recommends meeting the phase-out deadlines established by the federal government and the development of an education program implemented through extension services to inform and promote safe alternatives to the toxic chemical. It also recommends protecting farmworkers who use the poison and launching a program to reduce pesticide use in Florida overall by promoting sustainable pest management. This method, also called integrated pest management (IPM), uses methods of pest control that involve no toxic chemicals. Instead, IPM uses harmless products such as neem, a powerful bug repellent made from the bark and oil of the neem tree, or specific methods of planting that discourage disease and pest infestation. Other methods include covering plants with mesh fabric to prevent insect infestation or removing bugs by hand.[9]

In the report, the group also considered methyl bromide a public health hazard for people living near the farms that use the fumigant. After studying several Florida communities, it found three counties where fumigated fields create concern because of their proximity to local schools. Gadsden County, in the panhandle, has 2,950 acres of tomato fields treated with approximately 395,000 pounds of methyl bromide annually. Gretna, a small Gadsden County town of about 2,300 people, sits in the middle of these treated fields, exposing children attending nearby Gretna Elementary School to the chemical.[10]

Dr. Marion Moses, director of the Pesticide Education Center in California, works closely with the United Farmworkers assessing the many pesticide risks threatening agricultural workers and their families. Since many farmworkers often live close to the fields, they are exposed to dangerous petrochemicals even when they are not working. The land their children play on is toxic, the water they drink may be toxic, and they must handle the toxic chemicals at work.[11]

Though Moses concedes that the chemical poisoning is less acute than twenty years ago, growers continue to use new and still dangerous chemicals. Parathion, a toxic nerve gas developed by the Germans during World War II, was discontinued because of farmworker protests. It came on the market in 1943 and was responsible for more poisonings and deaths than any other pesticide. But some of the newer approved chemicals may be just as toxic. The EPA rates chemicals from one to four, with one being the most toxic. Moses wants to see all "Tox Ones" (pesticides rated as one, or most toxic) removed permanently from use.[12]

Until not long ago, farmers allowed the crop dusters to spray the fields while the workers were picking, but after much protest the practice stopped. In 1990, environmental groups won a victory using the Delaney Clause, a 1958 amendment to the Food, Drug and Cosmetics Act. The Delaney Clause states that processed foods must not contain residues of any pesticides that induce cancer in laboratory animals. It covers thirty-six pesticides including mancozeb, a fungicide used on cereals and grapes; dicofol, an insecticide used on fruits; and captan, a fungicide used on plums, grapes, and tomatoes. Several California environmental

6.5. The cap this woman wears advertises a pesticide she probably works around frequently. Pesticide exposure is one of the major health problems field-workers face, but they are often unaware of the dangers. Farmworker advocates offer programs to teach proper handling that allows the least exposure.

6.6. Chemicals are used in all aspects of farmwork. Here a woman in a lean-to outside a packing shed wears gloves to protect her hands from the disinfectant in the water in which she washes the summer squash.

groups, farmworkers, and the state of California sued the EPA using the Delaney Clause, and a California court approved the settlement in 1995. Though this will lead to stricter regulations on the pesticides, a group called the American Crop Protection Association says the five-year review process only proves costly and time-consuming, with resources wasted "in pursuit of trivial and non-existent risks."[13]

Lake Apopka: Natural Wonder to Disaster

Florida is a watery state. Bordered on three sides by water, the state also has many wonderful lakes and rivers long recognized for excellent fishing and boating. Lake Apopka, just northwest of Orlando, used to be one of those beautiful lakes. It was known for its abundance of freshwater fish, and the marshy wetlands along its vegetation-lined shores were home to a huge population of alligators. It was so clear that fishermen used to say you could target the bass you wanted to catch. Now it is considered the most polluted lake in Florida, and it stands as lifeless testimony to what can happen to a pristine body of water when humans tamper with nature. After years of fish kills due to algae blooms, the death of hundreds of alligators in the mid-1980s, and most recently, the death of over 800 waterfowl, Florida is now taking extreme measures to restore the once-beautiful lake. Although each new measure seems to be met with a new problem, those concerned are striving to find answers.[14]

Environmentalists around the globe are following the sad story of Lake Apopka because it is a textbook case of a beautiful, living lake ruined in the name of improvement. The story is also an example of how the effects of pollution can change human lives and how costly they are to correct. As people try to reverse environmental damage that occurred over the last century in an attempt to manipulate nature, Lake Apopka takes on a new meaning as a bellwether for future restoration projects.

Lake Apopka began to die during Florida's land boom in the 1920s, when the lakeshore town of Winter Garden began dumping its sewage and wastewater from citrus processing plants into the lake. Farmers drained water from nearby muck farms into the lake, increasing the lake's nutrient load. But the worst degradation began in the 1940s, when large areas along the lake's edge were drained to allow farmers to utilize the rich muck stored in the lake bottom. During World War II, Americans were fearful there could be a nationwide food shortage, and the demand for farmland increased.[15]

"Before the muck farms began, Lake Apopka was the second-largest lake in Florida," said Jim Connor, who was project manager of the St. Johns Water Management District that oversees the lake. "Now it is the most polluted lake in Florida. During the 1940s, when muck farming began, and the farmers built levees and drainage canals, the lake went from 51,000 acres to 31,000 acres, so now it's the fourth-largest lake."[16]

No one at the time suspected the impact this drainage and constant farming would have on the lake and its inhabitants over the next fifty years. The area was named the "Zellwood Drainage District," and 19,000 acres of the muck land was deeded to farm interests. Farmers established their own rules and the right to irrigate their fields with Lake Apopka's waters. For about fifty years, farms discharged phosphorus into the lake, which led to deteriorating water quality in Lake Apopka and the adjoining Harris Chain of Lakes. The lake's decline is now blamed on the fertilizer and pesticide runoff from nearby farms and on the drainage that shrank its wetland edges.[17]

The toll has been heavy, not only on birds, fish, and alligators that live and feed from the lake but also on the people who worked the farms. It is a dicey problem, and emotionally charged for people on both sides of the issue. Some farmworker advocates in the area say as many as 5,000 people lost jobs when the farms closed down.

Lake Apopka first received worldwide attention in 1992, when biologists noticed that the alligators there were not developing properly. Their findings uncovered an unexpected problem: Many of the male alligators had increased levels of estrogen that caused undeveloped, infantile penises, while others were not properly gender-defined. Evidence pointed to Tower Chemical's spill of the toxic chemical dicofol in 1980 that killed 90 percent of the lake's alligator population. Dicofol, the chemical released by Tower, is related to DDT, a pesticide banned from use in the

6.7. Fields are sprayed with the herbicide paraquat after the last picking. This tractor mows the dead stubble in preparation for the next planting near Lake Apopka. For decades these poisons drained into the lake, making the once-pristine body of water a death trap for wildlife.

United States since the early 1970s. It is known to interfere with hormone activity.[18]

Jim Connor described how the farms worked. "The farms used gravity runoff," he explains. "Farms were actually below the level of the lake, so the farmers flooded the fields to kill nematodes, then pumped the water back in. This eventually led to a buildup of sediment to five feet on the bottom of the lake.

"Cleanup attempts failed because there was no holistic approach," Connor explains.[19]

The state finally decided the only thing to do was to buy out the muck farms that lined the north bank of the lake. Most of the growers were hard-working men who staked their claims just after World War II and had farmed the fertile soil since the 1940s. Now they owned small empires, operated by their sons and grandsons.

"Naturally, they didn't want to sell," said Connor. "The Zellwood Drainage and Water Control District included about twelve farms working together to cut the cost of pumping and levees."

The group was feisty, he explained, and the men balked at giving up their farms.

"The farmers finally agreed to a buyout because their cost to meet the regulations was too high. It's cheaper for us [St. John's River Water Management] and for the state to do the restoration."[20]

Between 1996 and 1998, the State of Florida bought about 90 percent of the muck farms around Lake Apopka. The federal government provided matching funds for the buyout. The total cost exceeded $100 million. In July 1998, the costly reclamation began, and the muck farms flooded. Standing on the dikes looking out over the shallow lake, one sees that the former vegetable rows remained visible in many places for over a year. Birds perched on bits of dead cornstalks poking out of the water, remnants of the famous Zellwood sweet corn. Farther out on the lake, hundreds of water birds feasted on the lake's fare. Pelicans swooped from above as wood storks and herons stood on long legs and trolled the bottom for a tidbit. Nearby an alligator draped itself on the side of a drainage ditch to absorb the sun.

6.8. Alfredo Baheña (*right*), pesticide coordinator for Farmworker Association of Florida, Pierson Office, talks with a fern cutter on a fern farm outside of Pierson. Many farmworkers displaced by the Lake Apopka cleanup effort found jobs at Pierson's fern farms. Compared to fieldwork in vegetables, fern cutting is a luxury job.

6.9. Ferns grow in wooded areas, and the modern farmers use shade tarps. But though work with ferns is cooler and cleaner than that in the fields, there are still pesticides to contend with, along with the same stooping and risk of repetitive stress injuries from constantly bunching the fronds in bundles secured with rubber bands. Workers earn from 19 to 20 cents per bundle.

With all of the humans gone and the farm machinery silent, the wildlife returned. But then, in November 1998, several months after the reflooding of the muck farms, more than 800 birds were found dead on Lake Apopka.

"Eight hundred birds have died, that we know of," said Connor, indicating that pesticide residue embedded in the muck may have been the problem.[21]

Some people involved with Lake Apopka's restoration find too little attention paid to the human element in this environmental drama. When the farms were bought out, many farmworkers found themselves out of a job after working in the local fields and packinghouses. Though there are some funds for retraining farm laborers for other jobs, some advocates say it is not enough, noting that far more went for the purchase of the reflooded farmland.

Jeannie Economos

Jeannie Economos is the former Lake Apopka project coordinator for the Farmworker Association of Florida. The group's office in Apopka is headquarters for the retraining project for farmworkers who have lost their jobs to the state- and federally funded muck farm buyout. She estimated that about 2,500 workers have lost their jobs, and this loss of work affects many thousands more when one considers the extended families of these farmworkers.

"The original buyout was expanded, and now the state has spent more than $100 million for the farmland, while legislation allotted a mere $200,000 for job retraining for the displaced farmworkers," she said. "Most of them have worked on farms all of their lives and have no other skills.

"Duda [A. Duda and Sons, a large international farm] was supposed to keep the packinghouse open for two years," said Economos, "but they closed it last week. It employed 160 to 170 people who came back to Apopka to work, but they now have no work.

"They have offered these people work in other farm areas of the state, such as Lake Placid, but they don't want to go there because there is no housing," she continues. "They'd have to rent their own places, and that's too expensive for most of these farmworkers. They don't have first and last month's rent and security deposits. They had trailers here, but now they have no work so they have to get their things out of the trailers and they have no place to go. The majority are people of color, all are low income, and a large percentage are immigrants. Approximately 40 percent are women."[22]

In addition to the lack of money provided for retraining these workers, little has been done to assess their health. The Farmworker Association of Florida held meetings in Apopka to address concerns of members of the community, specifically to talk about the news linking pesticide and bird deaths at the lake.

"We don't know who could be affected because health effects are so wide and so varied that unless you do a survey in depth, people say things are not related," said Economos. "We're hoping to get funding for a survey and use that data to do a large health study that would document some health problems. Lake Apopka is a very significant chapter in the study of endocrine-altering abilities of synthetic chemicals. We need the Department of Health to take a serious look at this, but they say the levels of chemicals are not significant enough to affect anyone."

Since the lake is the site of two Superfund sites (polluted industrial sites designated for cleanup by the EPA), Economos was also concerned about a small rural community near Zellwood that depends on well water." The EPA says local wells are not contaminated," she said, but she also expressed uncertainty about how thorough the tests were.[23]

So the battle to clean up Lake Apopka continues, and everyone watches. Around the United States, other cleanup and restoration projects are in the works. Whatever is to blame for the alligator deformities, the bird deaths, and the fish kills may not be corrected for years. The muck is now useless to both humans and wildlife. Now the state is treating the soil with alum to bind the phosphorus that causes the algae blooms into the muck so it can be removed. Some worry that treating a chemical problem with another chemical will only create more problems. In the meantime, Lake Apopka remains a complete disaster with no known remedy.[24]

Margie Lee Pitter

Margie Lee Pitter fidgets nervously as she sits in the Farmworker Ministry office in Apopka, where she has turned for help. She is one of an estimated 5,000 workers in Volusia, Putnam, and Lake Counties who have been put out of work by the Lake Apopka restoration project that closed the farms once lining the banks of what was once Florida's second-largest lake. Only Lake Okeechobee was larger.

After thirty-five years as a farmworker in the area, Pitter worked her last day in a carrot packinghouse on July 31, 1998. That was the day that farming along Lake Apopka's polluted banks officially ended and that the muck farming that was a way of life in Apopka ended with it. Since the farming stopped, Pitter has worked several different jobs, but none with which she has felt comfortable. Now, at the Farmworker Ministry Office, she is registering for a class to learn computer skills, something she hopes will provide enough income to keep her from having to receive public assistance.

Pitter's story is typical of many of Florida's farmworkers. She started working in the fields just out of high school, hoeing between the rows of chicory and endive. She worked in the thick muck, which was often dry and dusty and blew around, caking on her face.

"Sometimes we looked like a coal miner at the end of the day," she says. Pitter and her fellow workers wore skirts over long pants, because in those days there were no portable toilets available among the rows of greens. She worked those fields until 1989, when she transferred into the packinghouse. Now, after a bad year compounded by the end of the only work she knows, Pitter hopes things will get better. Last year her house was badly damaged by tornadoes that devastated areas of Central Florida, and then, in late November, her husband died, leaving her all alone.

"I never had no kids," she says in her soft voice. "I had four miscar-

6.10. Margie Lee Pitter sits quietly in the office of the Farmworkers Association in Apopka, waiting for computer training classes to begin. She is one of several thousand workers left without jobs since the state closed the farms polluting Lake Apopka.

riages but never had any kids. Just a few years ago a nurse with the health service came and talked to us about safety and health, and she asked if I had any kids. When I told her I had four miscarriages, she said that might be because of the pesticides in the fields. I never had any idea about that.

"When I started working in the 1960s, the boss just told us to go over to the side of the field when the plane came over with the spray so we wouldn't get sprayed on, but we all felt the mist in our faces 'cause it blew around. Sometimes we all had headaches, blurry vision and felt sick, but nobody told us it might be the chemicals."[25]

Pitter and thousands of others never knew the consequences of the pesticides sprayed overhead by the crop dusters. Most of those exposed to the poisons ignored the telltale signs of chemical poisoning. In those days before the law required farmers to tell workers what pesticides they were using, hazards involved many different chemicals, including DDT, which is now completely banned in the United States. Pesticides also can cause skin irritation from handling without gloves and lung damage from the fumes if no mask is used. There is an increased incidence of reproductive organ cancer among farmworker women at a much younger age than the average population, as well as a higher incidence of birth defects among their children.[26]

Now the restoration project is experiencing problems of its own. Land flooding stopped because of nearly 1,000 bird deaths believed to be related to eating contaminated fish from the lake. The restoration must go on, but the problems with the pesticides seem insurmountable.

Pesticides in American Fields

Pesticides pervade American agriculture. Most were introduced after World War II, when they were manufactured as nerve poisons. Many of these poisons are related to the deadly nerve gas sarin, which was used in Japan in a 1996 subway attack. That episode was an act of terrorism, but there are many reported cases of poisoning in every rural area in which farmers use pesticides.[27]

In November 1989, in the tiny southern Hillsborough County farming community of Balm, workers picking cauliflower began to feel dizzy and nauseated. Someone went to get the boss. By noon, many more workers were ill. Others who were able began taking sick workers to a local farmworker clinic in their cars. At the final count, 112 workers in all suffered poisoning from the pesticide Phosdrin (trade name for mevinphos), a powerful poison used to combat worms.

Goodson Farms, the workers' employer, had sprayed the fields just twenty-four hours before allowing the pickers to return to the fields. The recommended time to stay out of the fields was forty-eight hours. The powerful nerve gas penetrated the workers' skin, they inhaled it, and some who munched on the cauliflower in the fields consumed it. The Phosdrin episode became the largest case of farmworker poisoning in Florida, causing outrage among advocates who protested the lack of care for these workers. The workers sued Goodson Farms, but in the end each was awarded only about $1,000.

Phosdrin, or mevinphos, works by attaching itself to cholinesterase, making it inactive. When cholinesterase is not active, the body breaks down neurotransmitters, leaving the affected person with a lack of feeling. It also causes headaches, diarrhea, nausea, and stomach cramps, and in extreme cases, fills the lungs with secretions that can result in death. Some of these workers remained in unstable condition for months, and some suffered permanent damage and have never returned to the fields.[28]

In 1993, the EPA listed mevinphos as one of the most dangerous of twenty-eight chemicals, even outranking the highly toxic parathion. Though the EPA attempted to have an all-out recall of the chemical, foot-dragging by lawmakers allowed the chemical's use through the end of 1995.[29]

Children and Pesticides

Children living on or near farms face an increased risk of exposure to highly dangerous pesticides, according to a 1998 study by the Natural Resources Defense Council, a watchdog group that oversees environmental concerns. Children are especially vulnerable to the toxic effects of pesticides because of their small bodies and their habit of putting their hands and other objects in their mouths. Because their bodies and brains are developing, they are more susceptible than adults to toxins.[30]

The report, titled *Trouble on the Farm: Growing Up with Pesticides in Agricultural Communities,* highlights the problems of the more than 500,000 children under the age of six who live surrounded by these agricultural poisons. Residues from chemicals considered too toxic for domestic use show up in these homes, often brought in on the shoes and clothing of parents who work in the fields; the residues are found on the children's hands and in their urine. Often these residues are at levels exceeding the level currently considered safe. They may also be exposed through contaminated soil, through pesticide "drift" in the air, and through play near the fields. Atrazine, a powerful herbicide (recommended for use on St. Augustine grass, a perennial favorite for Florida lawns), was found inside all houses of Iowa farm families during the application season.[31]

Over fifty labor, health, and environmental organizations submitted an administrative petition to the EPA asking that farm children's safety be considered when designating "safe" levels of these chemicals. Under the Food Quality Protection Act of 1996, the EPA is required to consider children's special vulnerability when evaluating their exposure, requiring an extra tenfold margin of safety. However, levels of exposure are uncertain, and critics say the EPA's record in enforcing the law is poor. In 1991, an EPA study showed that 48 percent of children working in the fields were sprayed with pesticides at least one time.[32]

In the 1990s, a dramatic case of pesticide poisoning occurred in Utah that had repercussions as far away as Florida. It happened when a seventeen-year-old farmworker sprayed himself with a toxic chemical in a sprayer. He was hot in the fields, and he thought that the sprayer contained water. The boy died the next day of a brain hemorrhage, which authorities said might have been prevented had he had pesticide training. The death of the young man prompted the Pasco County Health Department to launch a pesticide education campaign targeting farmworkers and their children to prevent such a tragedy. There are about 10,000 migrant and seasonal farmworkers in the county, and the program is a joint effort between the county health department and Farmworkers Self-Help.[33]

Again, education is the key to prevent pesticide exposure of farmworkers. Today, posters—printed in both Spanish and English—that illustrate the dangers of pesticides hang in the clinics and advocacy offices. Environmental groups help workers ask for precautionary equipment and offer legal help if workers are exposed. Because children are most at risk from contact with these chemicals, they should not encounter them, either in the fields or brought home on the clothes of family members.

Pesticides pose not only a risk to humans but also cause economic concerns. The problems with Lake Apopka are like the canary in the coal mine. Toxic sites will continue to surface as people realize the dangers of these chemicals, and there may be many more people put out of work as environmental agencies reclaim these sites.

CHAPTER 7

Immigration

In 1993, Maria Enriqueta Quintero ran from a deputy she believed would deport her. A car struck and killed the frightened woman as she fled, darting onto Interstate 4 in eastern Hillsborough County. Deputy Given Garcia Jr. said at the time he had detained Quintero earlier because of a possible immigration problem but later released her without charges when federal officials declined to take her into custody. Family members maintained that $700 of Quintero's money vanished at the time. Her survivors filed a wrongful death suit claiming the deputy was responsible for her death.[1]

Quintero was one of thousands of farmworkers who come to the United States illegally, remain undocumented, and live in constant fear of immigration authorities. These workers may avoid going to help agencies because they fear deportation and separation from their families. Because of the workers' tenuous status, crew bosses may take advantage of these workers, knowing they will not report any egregious behavior to authorities.

Since the crackdown on immigration and the tightening of U.S. borders in 1986 resulting from the enactment of the Immigration Reform and Control Act (IRCA), many farmers claim that they cannot find enough workers to pick their produce. To find enough workers, some growers are pushing for the old practice of hiring offshore workers. The

7.1. This sign at Rainbow House, a building at Farmworkers Self-Help in Dade City, lists the services the agency provides, including help with immigration problems.

farmer makes a deal with a crew boss, or padrone, who brings the approved temporary workers into the country, rents them to the farmer, and takes all responsibility for them. Though the farmer pays the crew boss for the labor, the workers may leave the country with next to nothing after the boss extracts what he figures they "owe" him.[2]

Farmworker advocates estimate that from 60 to 70 percent of Florida's workforce is illegal. In Buffalo, New York, several United States Immigration Naturalization Service (INS) workers were accused of harassing farmworkers picking the cabbage crop. According to the *Buffalo News,* in November 1997 a farmworker reported that INS agents chased several of his coworkers into a cornfield during an arrest at about 6:45 A.M. The worker said he and others were handcuffed, put into leg restraints, and treated in a rough manner by INS agents. When some of the frightened workers ran into a cornfield, agents began chasing them with guns drawn, the worker said. Top officials of the Buffalo INS office said there have been hard feelings generated in recent months by the agency's crackdown on illegal aliens working on western New York farms.[3]

Immigration History

The 1986 Immigration Reform and Control Act (IRCA) was the most complete reform of the immigration laws in thirty-five years. Prior to 1986, the Immigration and Nationality Act of 1952 followed the old quota system model established in the 1920s, which regulated the number of immigrants from various countries in favor of those with job skills and relatives of citizens or resident aliens. It also introduced a new wrinkle: a possible $2,000 fine or up to five years in prison for anyone knowingly transporting or hiring undocumented aliens. However, though it punished illegal aliens with deportation for working in this country, it did not prevent employers from hiring these undocumented workers. This loophole allowed exploitation of many undocumented foreign workers. If they complained about working conditions, they risked deportation by their employer.[4]

7.2. Haitians picking snap beans in a South Florida field. Until the Haitian Immigration Act of 2000, there had been no amnesty program for Haitians since 1987. The only way they could get a green card previously was to work ten years or marry an American citizen, said Flovil Samedi of the Haitian Refugee Center in Miami.

The first temporary foreign workers legally brought into the country were the *braceros*, mostly Mexican men, who labored for a specified period of time and then returned to their own country. *Bracero* is Spanish for arm, making the term roughly equivalent to the American term *hired hand*. During the Second World War, American fear of not having enough food fueled a drive by growers to import these foreign workers to prevent a shortage of farm labor. East Coast farmers feared domestic workers would not be able to travel along the regular route for harvests because of tire and gas rationing due to the war. In 1942, the U.S. government began the *bracero* program, allowing farmers to bring in foreign workers on a temporary, as-needed basis.

By 1943, 4,000 Mexican workers were harvesting in California and Arizona. The Farm Security Administration housed these men in migrant farm labor camps. In North Carolina, tobacco producers asked to double the number of men in the labor camps. When African American domestic workers held out for higher wages, growers called them lazy. The Mexicans living in the labor camps were managed more easily. In Clewiston, bosses at U.S. Sugar complained that their African American workers were leaving and going to pick beans. No wonder, when bean pickers could earn in a few hours the same amount that cane cutters earned in their ten-hour workday. For growers, importing workers was good business because these workers were unable to leave the cane fields to find more lucrative work.[5]

H2A Workers

The *bracero* program ended in 1964 due to protests by labor unions and other groups concerned about the abuses as well as the hiring of foreign workers in place of domestic workers who needed employment. A new program for importing offshore workers, the H2 program, replaced it.

During the time it was in place, the *bracero* program brought an estimated four to five million Mexican workers into the United States and set the pattern for illegal immigration from Mexico. When the *braceros* returned home from working in the United States, they had money, which made others want to come to America to work. "H2" became the name for temporary offshore workers, and the sugarcane industry around Lake Okeechobee imported about 10,000 West Indian "H2s" each year to harvest the cane. Now cane cutters are no longer needed, because in 1996 the industry introduced machines for cutting the cane.[6]

Florida farmworkers who have obtained citizenship or have earned their green cards are worried about losing their jobs to temporary guest workers from Mexico, Jamaica, and even such far away places as China. Under provision H-2A of the 1986 Immigration Reform and Control Act, Florida's growers can now access a huge Third World labor force under certain conditions. Farmworker advocates see provision H-2A as a method of maintaining a cheap and unending supply of labor. By law, any grower who wants H2A workers must prove that there are not enough domestic workers available to meet the demands for wages and conditions set by the Department of Labor. In Florida, that means at least $5.15 an hour, free accommodations, free transportation, and three meals a day. Farmworker advocates say the strict regulations governing all aspects of the H2A program are a sham, because state and federal governments cannot or do not monitor what happens to workers in the closed camps.[7]

Fernando Cuevas Sr.

Fernando Cuevas Sr. is national vice president of the Farmworker Labor Organizing Committee (FLOC). In addition to his concerns with farmworkers, Cuevas also goes to Mexico to inspect the conditions in the *maquilladores,* the Mexican and Central American factories that make so much of the apparel sold in the huge chains of discount department stores that pop up everywhere in America.

Cuevas was born in Brownsville, Texas, into a family of migrant farmworkers. His family followed seasonal crops around the United States, eventually settling into the eastern migrant stream. Cuevas remembers the traveling, the never-ending labor, the poverty, and especially the abuse he witnessed toward farmworkers. He was only a child of five when he saw an angry boss kick his father in the rear for no particular reason.

"I said, right then, I will become a crew boss so I can treat the workers better," Cuevas says. "I was born in Brownsville and became an interpreter for the family because I went to school and could speak good English. We traveled in the back of a two-ton truck with canvas over the top. The *troquero*, or *contractista*—the guy who owned the truck and transported us—would take the canvas off the truck when we got where we were going and use the same truck for hauling the produce. We would go through San Antonio and get an advance from the sugar beet company and then go to Ohio to pick the beets. I worked under my grandmother's Social Security number when I was underage. That was when you had to be sixteen to get a Social Security card. Then I worked under my mother's number. The contractor makes only one check, so the wife and kids, if they work, are invisible.

"I was only eighteen when I became a crew leader to help the workers. I wanted to help my family, so I became a contractor to make more money. Then I got into organizing, so I got rid of everything. I decided I couldn't own anything if you're an organizer, because other workers would be jealous. I struck against Campbell Soup in 1978, and I changed. I wanted to help. I was hired by the Farmworker Ministry so I could have an income while I organized—they gave me a salary. In 1975, I bought a two-bedroom home, for nine people.

"North Carolina brings in the most H2A workers. About 24,000 are brought into the U.S [annually], and 10,000 go to North Carolina—that's half of all workers brought into the country. The North Carolina Growers Association has worked with the Florida Fruit and Vegetable Association to obtain H2 workers in Florida. We call the recruiters 'coyotes.' They are bilingual people who bring the H2 workers to the location where they are picked up by the transportation arranged by the North Carolina Growers Association.[8]

"Many workers who travel with their families will move to another location just to get housing. If they can secure housing, sometimes they will stay even after the season and do work around the area just to keep the housing. Certain families do only hoeing; some work sweet potatoes and tobacco and then move on to other areas, north to Ohio, for example.

"Growers say they need the [H2] workers from April to October, so even if they have no work they use them to repair barns or something. It's like slave labor. They keep them here for nine months out of the year. Though they say they pay minimum wage, there is no enforcement. The payroll is padded. When I was a crew boss, they told me that the hours to claim they worked must match the payroll. They would pay for each worker's piecework, but if the piecework didn't total minimum wage for hours worked, we had to lower the hours or get fired. The pay for piecework is the same as twenty years ago."

Chemicals pose the most dangers to farmworkers, says Cuevas, echoing the sentiments of most advocates in Florida.

"Anywhere they are using chemicals is dangerous. They say they educate the workers, but many workers don't speak any English, so how can they read the warnings or understand the directions for safe handling? I was thirty-six years old before I heard the word *pesticide*. Most migrant workers have no Workman's Comp, so if they are injured they must pay for health care themselves.

"The Worker Protection Act of '95 states workers must be trained to handle pesticides, but all they do is show them a video. The problem with the H2 workers is many don't even speak Spanish—they often speak their native dialect. Many are Azteca and speak that dialect. No one can understand them here in the U.S., so how are they supposed to understand about pesticides? In North Carolina, the growers give workers the number of the North Carolina Growers Association to call in case of poisonings. Why don't they give them a number for poison control or the local emergency room? But where would the workers call? There aren't phones in the fields, so they wait until they are where they can get help, but then the exposure is long. Because they often don't understand the language or what pesticides can do, they don't know the symptoms or how they should wash up and change their clothes."[9]

Sister Teresa of Indiantown

Indiantown, just ten miles east of Lake Okeechobee near Florida's east coast, is a tiny town that took its name from a Seminole campground once on the site. It is where many Guatemalans settled beginning in 1982, escaping the political violence in their homeland. The Guatemalan families arrived speaking the dialect of the Kanjobal Indians, descendants of the Mayan civilization. Father Frank O'Loughlin, parish priest of Holy Cross Church in Indiantown and a seasoned farmworker advocate, let some of them stay in the church for a few days, because they had no place else to stay. In 1983, INS arrested seven Guatemalans, but since no one could speak their language, the authorities listed them as John Does. INS judges, horrified by the violence in the Kanjobal villages, allowed the Guatemalans to be released in the care of Father Frank's parish. Although Father Frank is now in Boynton Beach at another parish, the Guatemalans are firmly established in Indiantown. Currently their advocate is Sister Teresa.[10]

Sister Teresa, a Dominican sister originally from Bolivia, has been in this country for thirty-five years. Eighteen of those years have been in Indiantown, where about 3,000 Guatemalans make their homes. She is an advocate not only for the Guatemalans but also for the Haitian, Jamaican, and Mexican farmworkers that live there. She talked about the problems of the Guatemalans, who had an especially difficult time when they first arrived in 1982. At InDios Cooperative, Inc., a worker-owned sewing organization, Sister Teresa oversees the operation.

"We make shirts for priests that we sell all over the country. There are eight employees, and we train them. The first Guatemalans arrived in 1982. There were forty of them. I remember exactly because that was my first year here. The crew bosses went to the border of Mexico and picked up the men. Because they were small, the bosses thought they would be able to climb the ladder easily to pick the oranges, but they didn't have enough strength to lift the sacks—they weigh ninety pounds. The Guatemalans are small people with small frames. So they went to the fields to pick cucumbers and beans.

"The first group to arrive were Kanjobal Indians. Now we have many Guatemalans from all parts of the country. They speak thirty-two different dialects, because each region has its own. It was hard for them in the beginning, because they didn't know anything about America. Now we have at least fifty Guatemalans who are citizens, but there are about 3,000 here in Indiantown.

"Indiantown is like a sanctuary because of the large Guatemalan population. INS doesn't have raids here like they do in El Paso and the border towns. They know they are political refugees, and they can't send them back because of the terrible troubles in Guatemala. President Clinton knows that they can't go back, so there is no pressure from INS. We have trained our people to ask for a trial. If they are stopped by immigration, we teach them not to say a word but to ask for a lawyer. They don't answer any questions. That way they go to trial. Since the government doesn't want to be involved in a trial, they don't bother the Guatemalans. Other farmworkers—the Haitians, Jamaicans, and Mexicans—come here because they don't get bothered by INS much here because Indiantown is left alone.

"There is a policy for the Guatemalans called NICARA (Nicaraguan Adjustment and Relief Act) to apply for citizenship if they can prove they have been here since 1990, but there is no deadline for the application, because they are refugees. We hired a Guatemalan woman who is trilingual who's doing the initial work with them. There has been no official government statement on the Guatemalans in fifteen years, and they have no deadline for applying for citizenship. The Haitians must apply by March 31, 2000. Haitians that can prove they have been here since 1990 can apply for citizenship, but for the Mexicans, there is no way they can become legal in this country. They are the largest population here in Indiantown, but if they are stopped, they say they are Guatemalan and they are usually left alone. The Haitians can't do that because their skin is darker than the others, so it's a racist policy. They can be sent back to Haiti [if they are undocumented]. The INS doesn't see them as political refugees.

"Now most of the Guatemalans work in landscaping, nurseries, in restaurants and hotels. They work as waiters in the Chinese restaurants. They say they are Chinese, and that works.

"I do social work for anyone who is poor—Jamaicans, Haitians, Guatemalans, Mexicans, whites—I don't differentiate. I do some real estate helping them purchase homes. Sometimes I do refinancing, but mostly I help them buy primary homes. I also help with accounts, and I talk to the banks for them. And I help with the school system. I also do citizenship classes for Martin County. Because I only charge $20, I have people come from all over—as far away as Jacksonville—to study here.

"There is no crime among the Guatemalans; they are very passive people and live in very close families. They do drink, and American beer is double [in alcohol content] of that in Guatemala, so one beer is like two. A six-pack is like twelve in Guatemala. That's not too helpful. Most of them just drink a little, have one beer after work, but a few, the heavy drinkers, are here all alone.

"But they are victims. They are not accustomed to banks, so they keep their money in their shoes. That makes them targets. Right now I'm teaching a course in Stuart about money and teaching them to use banks. One bank is working with us helping them set up accounts. They are also learning English, if not in adult education classes, from their company. The citrus farms, chicken farms, the tile company, they have found it's an incentive for their workers. The Guatemalans are very good workers and the companies like them."[11]

Maureen Kelleher

In 2000, the INS established the Haitian Refugee act known as HRIFA, which allowed amnesty to Haitian workers who could prove that they had been here for five years. As the March 31, 2000, deadline loomed, Haitians swamped immigration lawyers to help them establish the proof they needed to participate in the amnesty. In Immokalee, where Haitians are a large part of the farmworker population, Maureen Kelleher, an immigration lawyer, of Immigration Advocacy Center had her hands full. Hundreds of Haitians who work in the area came to her for help seeking citizenship.

"We have a heightened practice at Immokalee," Kelleher explained. "Now it's Haitians, but a few years ago it was Mexicans everywhere. The Haitians must file by March 31, so my hands are full until then. Now they're coming in bringing bags of paperwork to prove they've been here since 1995. They must show completed forms documenting they have worked ninety days or more. They must show rent receipts, electric bills in their names, phone bills, car payments—anything to offer proof.

"Because the education is not that great in Haiti, many of them are illiterate, so they come in with little chits of paper and all sorts of things. We work with them. They also must provide a National Archive birth certificate from Haiti. Many of these Haitians came in on 'parole,' meaning they were brought in for a special purpose.

"There may be some around, but I don't know of any who jumped the cane," she said, referring to the Haitian sugarcane workers who came in as H2s before the mechanization of the cane-cutting operation. Kelleher says that farm work is exploitation. "It's subtle. If you don't know the language, you don't want to make waves."

She was also displeased with Senator Bob Graham's proposal for a farmworker amnesty. In January 2000, Graham proposed legislation that would allow farmworkers to get a green card in five years if they work 180 days a year and get on a register.

"The farmers keep the register. That means the farmer's in control of your documents. Now, anyone who knows farmworkers knows that they don't always return to the same farm, so it would be hard for them to keep track of these documents. It really isn't an amnesty at all—it's a Trojan horse. And if they want the farmer to sign the documents, they must docilely accept their pay."[12]

Belle Glade: "Her Soil Is Her Fortune"

On the southeast edge of Lake Okeechobee sits the small town of Belle Glade, perhaps the nadir of the eastern migrant stream. Belle Glade is the quintessential farmworker community, located in one of the richest areas of soil in the world. It served as the backdrop for Edward R. Murrow's landmark 1960 documentary, "Harvest of Shame." Many of the tenements photographed in that film still stand, housing farmworkers now as they did then. The loading dock near the center of town still serves as the gathering place for farmworkers in the predawn hours, when the trucks come to take them to the picking fields. The ritual is still much the same, with the workers looking for the best deal, the first pick, and the best conditions. Now those who want to work in the fields are bused as far away as Apopka for a day's work. Big Sugar has become so big that sugarcane growers took over the vegetable fields, leaving only the surrounding sea of sugarcane that stretches for miles around Lake Okeechobee's southern shores. The offshore workers no longer come to Belle Glade to cut the cane, because the soil is stretched so thin over the limestone that the sugar companies can get mechanical cutting machines into the fields. Men are no longer needed.

For someone driving along the streets of the downtown community, Belle Glade has the feel of a Third World country. People of color fill the streets—African Americans, Haitians, Jamaicans, and others from the Caribbean. Buildings are painted intense blues and pinks, and trash litters the streets. Women sit in chairs at the doors of their tenement buildings hollering at the children running back and forth across the narrow streets. The sounds of rap and reggae seep out of the many bars, and on corners are one-story stores with crudely painted signs advertising beer, wine, and lottery tickets. Some of the buildings are old two-story concrete structures with exterior walkways that resemble the old Florida motels of the 1940s. Most of the buildings have a few broken windows and sagging doors. The whole area appears to be forgotten by city officials, except for the constant presence of the prowling police car.

7.3. "Her Soil Is Her Future." At the edge of Belle Glade's city limits stands this sign boasting of the area's rich soil and flanked by the insignias of prominent civic organizations.

7.4. Belle Glade's workers still gather at the same spot they did fifty years ago. Although they call it the loading dock, it is only a large, barren parking lot where the buses come early in the morning to carry workers to the fields. Because there are few vegetables grown nearby since Big Sugar turned most of the rich soil into cane fields, these pickers may travel many miles on these buses to reach work. The terrain is so flat around Lake Okeechobee that the sky arches overhead like a blue and white bowl. (Photo by Nano Riley)

Dylan Morgan, a paralegal with the Migrant Farmworker Justice Project, says that if there are more than five families living in a building, landlords must provide each unit with running water. Most of the buildings have shared water and toilets, obviously lack air-conditioning, and are, at best, minimal housing. Inspections of these buildings are few, and inspectors often overlook infractions. The buildings are not farmworker housing provided by a grower but rentals owned by slumlords. Broken-down cars are a prominent feature in this bleak landscape, which is brightened only by a couple of murals. One depicts Malcolm X; another shows two black men with the slogan "He's Not Heavy" painted above them. Patches of garden brighten a few vacant lots where people grow tomatoes and small stands of sugarcane, and banana trees cluster in occasional patches. It is another culture, a world apart. The entire politics of the area is related to Big Sugar, and once again the farmworker community is out of work because of corporate economics.

In the 1920s, Belle Glade was labeled the "Winter Vegetable Capital of the World." When the earth froze over in the northern winters, the rich muck of Lake Okeechobee provided a soil that would grow anything. For farmworkers in the 1930s and 1940s, Belle Glade became a winter home. Traveling north in the summer to avoid the scorching heat of Florida, where few things grow from May to September save black-eyed peas, eggplant, and okra, the mostly African American farmworkers had a patterned life. Belle Glade was a thriving, though poor, community with steady work for these migrants, even if the pay was low.

Zora Neale Hurston described the Belle Glade of the 1920s in her novel *Their Eyes Were Watching God:* Tea Cake invites Janie to go with him "on de muck . . . down in de Everglades round Clewiston and Belle Glade where they raise all dat cane and string-beans and tomatuhs. Folks don't do nothin' down dere but make money and fun and foolishness." Hurston described the town in September, just before planting season, as a place brimming with excitement as hordes of workers arrived. Some came in wagons from Georgia, some in truckloads from all over. Families came with dogs, with all their belongings hanging from the broken-

7.5. Belle Glade's tenements date to the mid-twentieth century. Most of them are in disrepair, but inspectors seem to overlook many violations. (Photo by Nano Riley)

7.6. Haitian couples often travel and work together. This couple is picking beans in a field near Homestead.

down cars transporting them. She writes of the all-night jook joints and the hard work on the muck during the day.[13]

Year after year workers returned to Belle Glade each fall to work the muck. People established families, and some workers even bought their own small houses. Many rented the same places for years, with extended family members living in the same places year-round. That remained until Castro's revolution in Cuba sent many of the wealthy sugar barons north to Florida. Since sugar growing was what they knew, Lake Okeechobee became a prime area for growing more cane. Farm after farm converted from vegetables to sugarcane, which required specialized cane workers. The widespread abuse of the sugarcane workers is documented in Alec Wilkinson's notable book *Big Sugar* (New York: Alfred Knopf, 1989), which provides an in-depth look at the sugar industry and its practices.

Most of these workers came from Haiti and Jamaica as H2 workers. These men came into the country legally, slept in barracks, and lived much as slaves had in the nineteenth century. Stories of the cane workers go all the way back to the 1920s and 1930s, when crew bosses used to go to cities in the South and recruit poor blacks with promises of high-paying work and the great Florida weather. These hollow promises soon became nightmares as young men discovered the harsh work and the dismal living conditions. They stayed in crude dormitories, rose at four A.M., and were fed only as much as is necessary to keep them going in the cane fields. These men were forbidden to leave the compound, charged inflated prices for their travel to the area, charged for the tools they needed, and charged for their rent and food.[14]

In 1939, the sugar subsidy was instituted by Congress, guaranteeing a set market price for sugar that continues to this day. Because of this, American sugar is more expensive than any other sugar in the world. In the 1960s, sugar caused a great change in the landscape of the southern Okeechobee area. As more and more farmers saw the profits in cane, they gave up vegetable crops that were subject to disease and the effects of heavy rains or droughts. They saw no reason to grow tomatoes that can go bad or vary greatly in price when Uncle Sam guaranteed a set price for sugar, especially when U.S. Sugar, one of the largest sugar producers, would buy everything they produced and set them up with the chemicals needed to bring in a good crop. Simple economics encouraged the smaller farmers to band together in cooperatives to cash in on Big Sugar's booty. But this left the pickers whose families had worked the vegetable fields for generations without jobs.[15]

Greg Schell

Greg Schell is a lawyer with the Migrant Farmworker Justice Project in Belle Glade. His record for winning suits for workers against unscrupulous bosses is renowned nationwide. Farmers know that when he is involved, they have a real fight on their hands. After graduating from Harvard Law School, Schell decided to come to Belle Glade because he wanted to do something meaningful. Belle Glade has been his home for twenty years. He talks about the local farmworkers, both citizens and undocumented workers.

"Now workers can travel north in the summer, but in the winter they now have to go miles away, north to Apopka or southwest to Immokalee, journeys that require rising early to travel to the picking destination in crowded, rickety school buses. Since the majority of Belle Glade's residents were African American, when they allowed farmworkers to get unemployment, that is what these out-of-work pickers did. They collected unemployment."

Schell talks about Big Sugar and the cane cutters.

"It's all labor economics. It costs more now to harvest by hand. It went up dramatically in the early '90s, when the wages went up by law to $7.13 an hour for the foreign workers. That's when they went mechanical. They are required to pay both foreign and domestic workers the same wage, but they usually don't. Some of the companies went to electronic timekeeping. They found the piece rate was no longer sufficient to meet the hourly wage rate. Now they use machines, not people."

Advocates such as Greg Schell were watchdogs for the workers, requiring the companies to pay by check rather than in cash. This eliminated the long-time practice of paying far less than the time a worker actually spends laboring.

"When advocates for the sugarcane cutters convinced the government to require electronic time clocks to protect the workers, then the growers decided to switch to mechanical harvesters," says Schell. "Though these machines had been available for over twenty years, the cane growers didn't use them because they were expensive. When the growers were forced to pay the workers the $7.00 per hour, they decided it was cheaper to use machines."

Another factor leading to the machine harvesting of cane is that the thick muck has slowly disappeared in many places, leaving barely an inch of soil over the hard limestone in many places. (This has made possible the use of heavy harvesting machines that earlier, when the dirt was several inches thick, would become mired in the thick muck.)

What distinguishes Belle Glade from other Florida farmworker communities is the population. There are fewer Hispanics in this town on the Lake Okeechobee shore than in the other rural farming communities in Florida. Most of the families are African Americans, Haitians, and other people of color from the Caribbean.

"Many of them are U.S. citizens," says Schell. "These families do not migrate up the stream as they did twenty years ago. Many of them still live in the old 'colored' community built in the 1920s to house the migrants."

According to Schell, many of the farmworkers are now able to obtain unemployment during the summer heat, when there is no harvest. Many who would like to work are being pushed out of farmwork by the influx of single, undocumented men from Mexico because bosses prefer them.

"Today's large corporate farms like these single men who come without families in tow and do not require much in the way of housing. These men pile up in dormitorylike buildings or live several to an apartment," says Schell. "Providing for the often large, extended Hispanic families that often include grandparents is difficult, and children attract the authorities. And housing for a family is more difficult to provide than housing for a group of single men, who don't mind cramming in dormitory-style for a few months while they have a chance to earn enough money to send home or buy a car.

"Some of the small family farms are still there who hire families. Those small farms are rapidly going out of business or are being bought up by larger operations. Most farms prefer the single men because they are less demanding, and since they only expect to work for a few years, they aren't interested in organizing. We have an increasingly alien workforce of single men from Mexico who want to get into construction or service jobs."

According to Schell, the prevailing model used by most advocacy groups, that of helping the migrant family, is no longer applicable. Schell believes there needs to be a new awareness of these migrant men who travel alone and are often in the country illegally.

"They want to remain as invisible as possible, so they don't complain about the work," says Schell. "Workers' rights and increased benefits don't interest them, because they expect to get better jobs in a year or two."

Schell echoes other farmworker advocates, blaming low wages and poor working conditions as the root of many other problems for agricultural laborers.

"Here in Belle Glade there are families living in tenements built for men, with crowded conditions and shared toilets. The blacks were confined to that area in the '20s, and it remains today. The two housing developments on the edge of town were originally migrant labor camps built in the 1940s and dedicated by Eleanor Roosevelt. They've never been incorporated into the town because the residents are black. If they incorporated those two camps, the town's population would be predominately black, and the whites don't want that.

"We need a limit on new workers coming into the country. There are at least two workers for each job. Here in Belle Glade most of the population is legal—African Americans and some legal Haitians—and they are not getting the jobs.

"Right now, almost all of the strawberry workers in the U.S. are either H2A workers or they are illegals," says Schell. "Florida is lagging behind

7.7. Greg Schell, a lawyer for the Migrant Farmworker Justice Project in Belle Glade, takes legal papers to a group of Haitian men who sued growers to get back wages owed them. The patch of banana trees on the vacant lot in the background looks a bit like their homeland.

7.8. This mother sneaked across the border into Texas with her oldest son when he was one year old and she was pregnant with her second child. A month later, INS deported her to Mexico City, where she stayed with relatives until she sneaked across the border again a month later. Now she is expecting her third child. Should she be returned to Mexico again, her children will make it more difficult for her to sneak back over the border. Her need for refuge may also put her at risk for abuse.

in this trend because it has such a large indigenous population of farmworkers—Hispanic, African-American, and Haitian.

"These [illegal] workers may show you something, but it is probably not a legitimate document," says Schell, referring to the papers carried by many illegal workers. "Documented workers complain, so it is preferable to hire illegals, which creates a chronic surplus labor force. Now many locals can only get a job at the height of the season."[16]

After George W. Bush took office in 2000, Vicente Fox, Mexico's new president, visited the United States and increased the pressure to allow amnesty for undocumented Mexican workers already in the United States. But since the terrorist attacks on September 11, 2001, America's heightened xenophobia has dampened any plans to increase foreign workers in the country.

"For the last six or seven years the agricultural community has wanted relaxation of the rules governing the use of guest workers, primarily because several features of the current law bother them," says Schell. "Growers view the current law as a barrier, because they must provide free housing for the workers. Most farmers in base states—states where migrant workers spend most of the year like Florida, Arizona, and California—don't have housing in place. For them to hire H2A workers means they must build housing, and that's a big, up-front capital expenditure.

"Second, they're concerned with the wage rate they must pay under the current program, which must meet state standards and is sometimes a substantial increase in what they normally pay domestic workers. With the H2A workers, their pay is a matter of record."

Prior to September 11, there was a compromise bill on the table that pleased both growers and farmworkers, but now much of the carefully crafted immigration legislation for foreign guest workers is on hold.

"It's now a question of whether growers are willing to spend their limited amount of capital to get a bill through a Congress that is extraordinarily concerned with border security," says Schell. "Congress is generally in an anti-alien mode, which is a switch from where it was just before September 11.

"There's always been a problem with the guest worker program," explains Schell. "There are never as many positions as there are workers who want to come into this country. That's always been the situation with the Jamaicans, where we have had a guest worker program for fifty-plus years. Growers want men age forty or less. They do a police check, a health check, and they're allowed to pick and choose the very best. And there are no laws that prohibit that. Recently there was an age discrimination suit on behalf of a Mexican who had come as a guest worker but suddenly found himself barred when he turned forty. The U.S. Court of Appeals for the 4th circuit, which covers the Carolinas, concluded that our discrimination laws don't apply, so that's one of the problems. We want the same standards to apply to guest workers that apply to U.S. citizens. In this case, the courts said that sort of discrimination is permissible when you're talking about bringing workers from abroad to come and work in this country."

Schell explains a consensus bill for guest workers, reached in October 2000, to which both growers and workers agreed. "Our bill is legalization with an earned component, where the worker who had done a small amount of farmwork could earn a temporary card that would convert to a permanent resident card after an additional few years of farmwork. Then they would be free to go and work at whatever they wanted," Schell says, explaining the legislation crafted by the United Farmworkers and other advocates. "Farmworkers will not sign off on a bill that doesn't have a legalization component. The compromise bill was essentially some modifications of the current guest worker bill in exchange for a widespread legalization program for current farmworkers as well as additional labor protections. Everybody agreed to it, but the strategy was to wait until after the elections of 2000 and add it to a 'must pass' spending bill. Trent Lott agreed, but Phil Gramm said it would go through over his dead body, so that shut down the whole operation."

Currently, with the increased border patrols, some growers are lamenting the lack of workers coming into the country. And Schell says that current anti-immigrant feeling makes farmworker legislation advocating any sort of amnesty for foreign workers out of the question.[17]

But no matter what the laws, immigrants will continue to make their way into America, providing a constant stream of new workers for agriculture. These marginalized workers create an underclass in America as they toil at jobs most of our citizens refuse. These new demographics show agricultural labor is a workforce in continual transition.

CHAPTER 8

Family Life

Most farmworkers are family people, whether they travel or work locally. Perhaps the isolation of being foreigners in a different culture creates a sounder bond, but they usually live in extended families, working together and helping each other in everyday family life. Though sometimes marginalized or stereotyped as irresponsible, most farm laborers are hardworking people trying to make ends meet. Even those young men traveling alone harbor dreams of bringing families to America, and they usually send money back home to help their families. Parents encourage children to get an education in the hope that they will move out of farmwork into a more stable and better-paying occupation.

Day care is an important consideration for all working women, but for farmworkers it may be unaffordable or difficult to find. If there is no place for preschool children, parents may take them to work, where they spend the day playing along the edge of the fields. Usually a farmworker advocate group sponsors some form of day care for toddlers too young for school, and the Migrant Head Start program is available in most communities with large farmworker populations. The Head Start program is valuable in many ways, because the Head Start educators encourage parents to become involved in their children's education. It also offers parenting advice as well, including health and nutrition information.

Mary L. Martinez

Mary L. Martinez is the education coordinator and assistant director of the Hacienda Head Start program at Camp Hacienda, a migrant camp outside of Naples. She has worked at Hacienda for twelve years, but before she worked with Head Start, she and her family were farmworkers. She understands the value of Head Start, especially in providing some day care for little ones too young to attend public school, and she sees the families every day going about their lives.

"We are right in the Hacienda migrant camp, which is mostly trailer houses. We are the only building that's not a trailer. The building is dedicated to Head Start. We have four classrooms covering birth to one year, one to two years, two to three, and three to five years. Then they go to public school. It's like a day care center, but we do lesson plans with the little ones and prepare the preschoolers to go to public school. We take field trips to the schools so they know what to expect.

"We have forty-eight children now, but we have the capacity for sixty. It changes seasonally as workers migrate. Our priority is migrant children. We take seasonal workers' children for two years, but after that they have to go to public schools. We're part of East Coast Migrant Services, so we do teach ESOL classes, plan to do GED, and we have parent meetings. We bring speakers to talk about nutrition, AIDS, and other health issues. When the children have a health problem, we give them a referral to see a contract physician we work with. We must write the child a note saying the child has green nasal discharge. We can't just write the child has a cold because they won't take them. Then we instruct the parents about how to give any medication they get and make sure they bring it with them to school. We also try to make sure the parents give them the right dosage at home, and we do it in the daytime.

"There are about thirty-eight families here at Hacienda. Most are Mexican, but we have one Cuban family, two from Puerto Rico, and four or five Guatemalan families. We also have young single mothers—some as young as seventeen. This year there are three young women with children. Most of the families have two to three children. We used to have a single father with six kids. He came for several years, but he has not been here for a while.

"Most of the families are young—under forty-three. These families are close. We've never had any trouble or had to call the police about anything. This camp is run by Gargiulo Farms, and I believe the rent is $10 for each person in the household weekly."

Martinez talks about her life as a farmworker and how hard the work was.

"I worked with peppers, cucumbers, and tomatoes, okra and oranges. The okra was spiny, and the oranges are really heavy, but pulling plastic was the worst," she says, referring to the black plastic laid on the fields as mulch to prevent weeds.

"I hated pulling plastic. We started pulling the plastic in late May. First you cut the middle, pull the plastic out, put it in a bundle, and carry it to a spot where everyone is dumping the plastic, and then go back and pull more plastic. It was really dirty. The dirt would go all over. It was right out in the hot sun, too. It was like an eight- or nine-hour walk, just walking and bending."

Martinez says she and her husband paid $40 per person per month for the trailer where they lived in the mid-1980s. Now she helps other farmworker families and their children.[1]

Helping Families

The Beth-El Mission in Wimauma, in southern Hillsborough County, is a busy help center for farmworkers. Their services and activities are geared to families, who gather there for holiday parties, church, and Sunday school, as well as English classes and GED courses taken to earn a high school equivalency diploma. Director Evan Jorn says the main purpose of the mission is to serve the large community of agricultural workers concentrated in southern Hillsborough County, and that means the mission always has something going on. Located between Parrish and Balm, it also attracts workers from nearby Dover, as well as Wimauma.

8.1. A father, still in his muddy work clothes, eats supper with his child at day's end.

8.2. Students from a nearby Presbyterian church in Sarasota join farmworker children at the Beth-El Mission in Wimauma for Bible lessons. The children and their families come to visit the mission several times a year to participate in services and to help out with volunteer programs such as the Christmas toy giveaway.

8.3. Father Ramiro Ros raises his hands as he leads the congregation at the Beth-El Mission in prayer. The service is evangelical Presbyterian, and the participants sing in both Spanish and English. Ushers pass out song sheets in both languages, and a projector flashes the words on the wall.

On Sundays, Father Ramiro Ros, the mission's pastor, and his wife, Olga, hold a service for prayer and worship at five o'clock. Afterward, the congregation always gathers for a church supper at long tables in a covered courtyard between the mission offices and the chapel.

Beth-El is evangelical, and during the service both Father Ramiro and Olga lead the congregation in gospel hymns sung in both English and Spanish. Copies of the words in both languages pass around the congregation while the words also flash on the wall. Families gather to worship, mothers holding infants while fathers manage the older children. The focal point of the chapel is the colorful stained-glass window high on the eastern wall that features a dark-skinned Christ with long black hair and Hispanic features. It is a joyful service. Some of the worshipers pray fervently, while others hug each other and cry. Large, brightly colored crepe-paper flowers, made by the Mexican women in the craft classes held weekly at the mission, decorate the altar.

In spring of 1999, one Sunday evening service featured a pastor from Honduras who showed slides taken during Hurricane Hugo. The pastor used both Spanish and English to explain the devastating aftermath shown in the slides. There were audible gasps from the congregation as he narrated in Spanish with Father Ramiro translating in English. There were pictures of people who had lost their homes living in cardboard boxes, picking through garbage for anything that could be of use. When the pastor asked for donations, everyone found something to put in the collection plate.

Taking part in the service on the evening of the Honduran minister's talk was a group of people from a Presbyterian church in nearby Sarasota. Everyone pitched in to help in the kitchen, preparing the church supper of spaghetti, salad, garlic bread, and pitchers of iced tea and Kool-Aid. Some of the workers who get assistance at the mission also help in the kitchen. There was plenty of food for all, even second helpings.

Olga Ros talks about the craft classes offered at the mission for women.

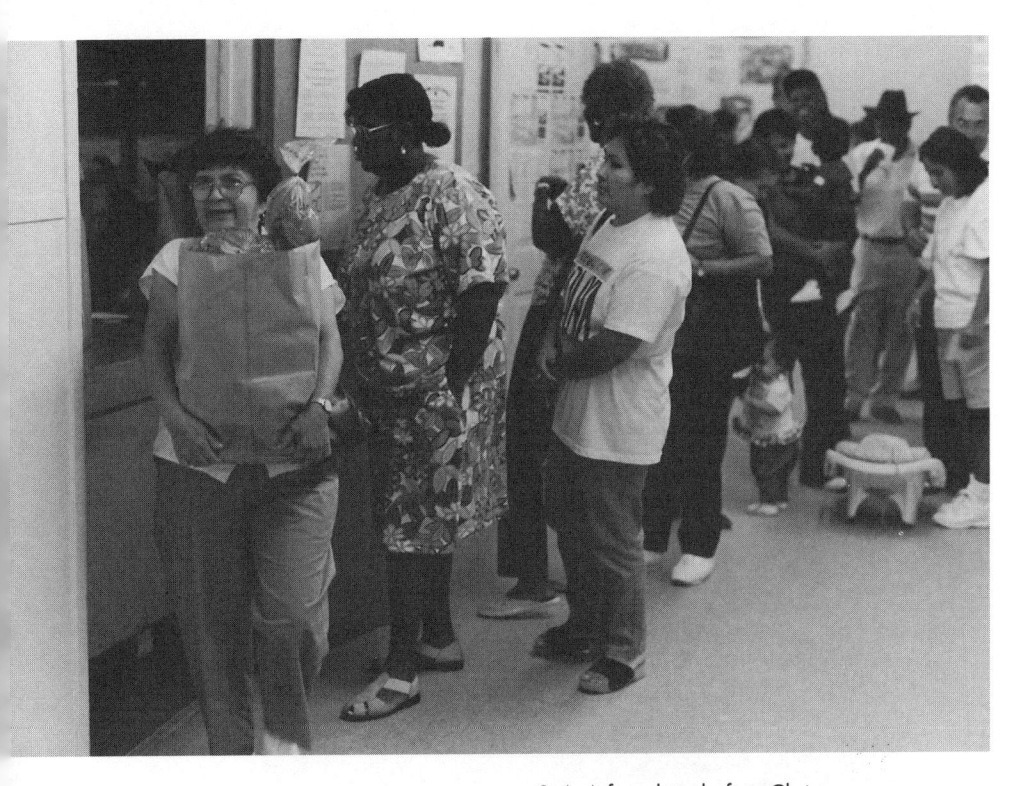

8.4. A few days before Christmas, farmworkers gather at the Beth-El Mission to collect their holiday groceries. In addition to such staples as rice and beans, they receive a chicken and a pie.

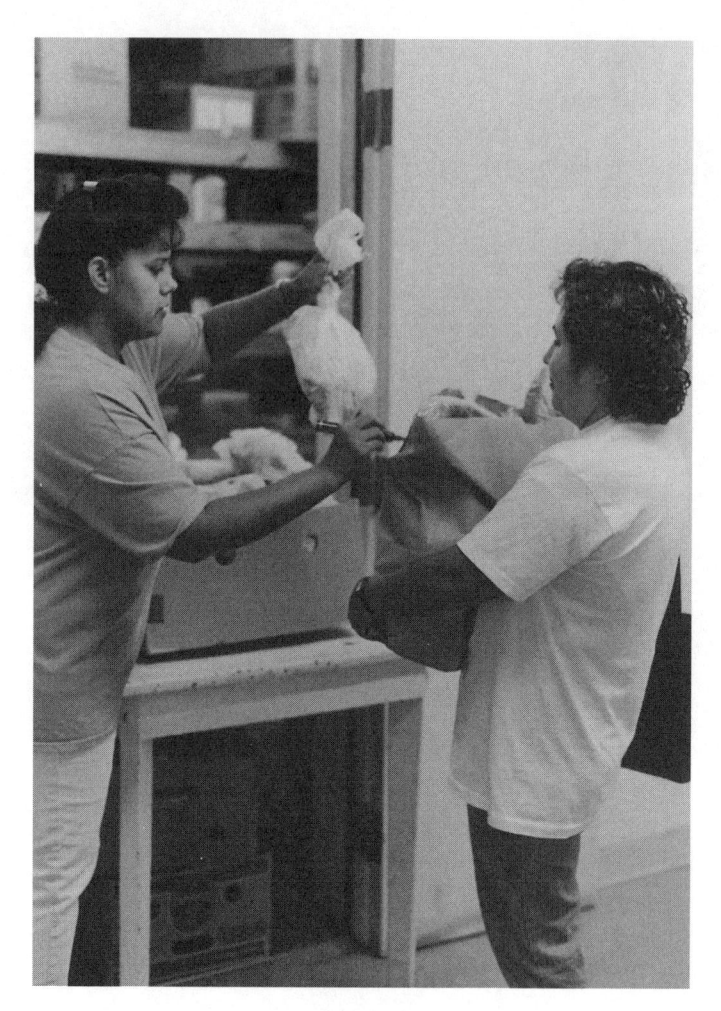

8.5. This woman receives her Christmas chicken at Beth-El Mission, along with other groceries. Local charity groups donate the food to the mission.

"We teach sewing as well as how to make the crepe-paper flowers that are traditional to Mexico," she says. "It provides socializing time. They're in air-conditioning, so they're comfortable.

"The women also make a delicious, traditional Mexican salsa which they sell to raise money for programs and events. During these sessions, we usually have a volunteer from The Spring [a domestic abuse shelter in Hillsborough County] who talks to the women. We encourage them to discuss their home lives, and many women for the first time understand they are victims of domestic violence."[2]

Gatherings for sewing and making crafts are popular pastimes for these hardworking women. In Dade City, Margerita Romo meets with a group of Mexican women for a quilt-making session on Monday nights at the headquarters of Farmworkers Self-Help. Here the women get together, swapping bits of material for quilts. Some of the children come with their mothers and grandmothers, sitting at nearby tables drawing pictures or chasing each other outside in the yard. These are pleasant times for the women, who share jokes and stories, laughing with two older American women from a local church who usually come, often bringing cookies and scraps of colorful cloth and other supplies for quilts. Some of the quilts become gifts or family keepsakes. Others make their way to fundraisers, where the maker and the sponsoring charity share in the profits.

These gatherings are important, because it is a time for the women to relax and chat with one another. Visitors often stop by during these evenings, bringing clothing or other donations that help out these families in their hardscrabble lives. A Pasco County health-care worker pays a social call every few weeks, and during her visit she answers various health questions and encourages the women to get checkups for themselves and their children. The familiarity puts these Mexican women at ease, making it easier for them to discuss family problems and take advantage of available help programs. Often if there is abuse at home, it is uncovered at these quilting bees among trustworthy friends.[3]

Though they work hard, these people find time to celebrate special occasions when they can have fun with all their relatives, young and old.

Birthdays and other holidays are an occasion for family parties. *Quinceñera* celebrations take place on a young girl's fifteenth birthday, the equivalent of a sweet sixteen party. Surrounded by family and friends, the young Mexican girls become the center of attention for the day. Families often spend several thousand dollars for the special party, which features catered food and dance bands after a special Catholic mass.

In many of Florida's rural communities, there is a growing Mexican presence. Workers gather at small bodegas that offer tacos and café con leche. The sounds of popular Mexican *norteña* music blares from the speakers behind the cash registers of these small stores. Nearby a rack of T-shirts for sale are emblazoned with Mexican symbols, including one of Emilio Zapata, the champion of indigenous Mexican people in the early 1900s. Many communities also sponsor festivals marking holidays celebrated long ago in Mexico. Even the single workers traveling alone come to these festivals, where they can have camaraderie and perhaps a taste of their homeland.

In 1999, Hispanic students at Manatee Technical Institute organized a welcome-back party for returning farmworkers in the area. In Immokalee, colorful posters advertise the annual Mexican Heritage Festival, which features traditional food and a mariachi dance band dressed in elaborate costumes and big sombreros. It is an occasion for friends and relatives of all ages to come together for a wonderful time and forget the hard work of the fields.

Epilogue

Farm laborers are the backbone of our agricultural society. For nearly four hundred years they have played an indispensable part in our food production. But at the beginning of the twenty-first century, it is important to realize that for farmworkers, little has changed. They still receive low wages that are inadequate for many of the basic needs most people in this country take for granted. The constant traveling drains both pocketbook and energy, sapping their hopes for entering the mainstream life of America.

As Americans become more health-conscious, demanding the freshest of fruits and vegetables, farmworkers continue to be an important and growing workforce. It is time to take a closer look at these valuable workers and to give them the respect they deserve in the form of better working conditions and higher wages.

From the time of the first agricultural workers in seventeenth-century America, their needs have been suppressed in favor of the growers' profit. The American philosophy of work contains the idea that employees of a farm or business learn their jobs and may one day go into business for themselves. This dream is not within reach of most farmworkers. For every farmworker who is able to stop traveling and enter another line of work, hundreds continue to labor in the fields with few guarantees of job or financial security. Theirs is a day-to-day existence centered on supporting family and home in the best way they know how. There are some changes for the better, but they often have been slow in coming and face much opposition from farmers and agribusiness along the way.

Though advocates fight for higher wages, the fact remains that bosses still want the cheapest labor. If they hire undocumented workers recruited by coyotes, they may pay them less than they pay local workers, putting locals out of work. Since it is impossible to completely seal the borders with Mexico, people still flow into this country willing to work for any wage.

In Florida, the unionization of the Quincy mushroom workers was a triumph for agricultural workers in this right-to-work state. Perhaps it will encourage unions in other areas of the state, though growers will fight it. The Quincy workers are under a microscope, and everyone in Florida's agricultural community is watching.

The traveling farmworkers must follow the crops, and this does not change. What does change are the faces of the workers. Many of the Mexicans who came with the *bracero* program stayed and became established. Now their children and grandchildren have moved out of migratory farm work into other jobs, and some have farms themselves, as Sylvia Medina's father does in Homestead, where he hires people to work in his fields.

In most farming areas in Florida, California, the Southwest, and the East, there are now permanent Mexican communities, with Mexican-owned stores catering to these workers. These *tiendas* sell the boots and gloves workers use in the fields and also stock Mexican foods, toiletries, and even the popular "novellas" (illustrated romance novels) and comic books. These stores are a focal point for the community, and the men stand around the door and share cold drinks after long days in the fields.

The newest farmworkers are the Guatemalans, Salvadorans, and Mexicans from the troubled southern states such as Chiapas, who are escaping the political upheaval in their homelands. These workers are taking the place of the Mexican families who have dominated farmwork since the 1950s.

In some instances, farmers realize it is necessary to provide amenities in order to have a satisfied and productive workforce. Marvin Brown and Jay Sizemore at JayMar Farms realized that. Their new, affordable housing will encourage many farmworkers to stay in the area year-round to work. Until more growers around the nation follow their example, many farmworkers will continue to live in dismal conditions, with nothing to encourage them to stop moving with the crops year after year.

There are still thousands of workers who are not so lucky. Those who speak little or no English remain at the mercy of crew leaders, and they are dependent on them for everything. These are the workers who are cheated and who live in the shabbiest housing.

Hundreds of these single men cross the border every year, sometimes legally, but most are illegal, the *sindocumentos*. These are the workers who will not organize or make any waves because they fear deportation. Often these undocumented workers disappear from an area a week or so before the harvest is finished, because they know INS workers may show up when the picking is over.

As long as growers and agribusiness continue to view these workers as mere laborers, there is little hope of improving their conditions. Faceless immigrants who move on after a few weeks show farmers nothing but their labor. Many workers are interesting and intelligent individuals, but few have a craft or skill other than fieldwork to allow them entry into the mainstream workforce in the modern United States.

The constant influx of these workers ensures that there is always someone to fill the empty space another agricultural worker leaves behind. All Americans bear the brunt of the farmworkers' problems, because most of the programs to help them are funded with U.S. dollars from taxpayers' pockets. Therefore, Americans would do well to pay more attention to improving the lot of these invisible workers who are so important to our lives.

Farmworkers seem to be caught in a time warp. Their wages are about the same as they were twenty years ago, which means that farmworkers are actually making less now than they did then. While the rest of the

United States marches forward into the technological age, farmworkers, especially the migrants, seem almost to move backward in time. They are dependent on the grower for work and on the government for benefits.

When agriculture entered the chemical age after World War II, using chemicals to control all pests and diseases, it also poisoned the land and water in many places. Now many farm areas slated for cleanup due to chemical contamination may be reclaimed by the state and federal government, as has occurred in Lake Apopka. These reclamation programs often leave many workers jobless. Such cleanup programs need to figure the human element into the equation and include retraining for the workers as part of the cleanup cost. Florida particularly needs to address this issue. As a state with a growing population and a shallow water table, more agricultural land may be taken out of use to preserve the water supply. Restoring the delicate balance of the Everglades will take more than just reflooding dry areas and removing drainage ditches.

Children rarely work in the fields as they once did, but there still need to be more protections in agricultural areas to prevent their exposure to farm hazards. Both the "Right to Know" law and the Farmworker Protection Act have made strides in helping workers understand the dangers of pesticides. New technology makes farm machinery safer, but not everyone has the latest equipment.

It has been more than forty years since Edward R. Murrow stirred America's conscience with his gritty television documentary on Thanksgiving night in 1960. "Harvest of Shame" prompted many valuable changes in the way people perceived farmworkers and their problems. It presented them as human beings with the same desires and fears harbored by all classes. It is up to the citizens of the twenty-first century to understand that it is not just the growers who cause the problems that farmworkers encounter. As long as people continue to eat the fruits and vegetables they provide for our tables, we are all responsible for their welfare. We are truly indebted to these valuable, invisible workers, for without their labor Americans would not have such an incredible bounty for their tables.

Notes

Prologue

1. "Study of the Migrant Labor Report," 1.
2. Ibid.
3. Ibid., 2–3.
4. *Migrant Farm Labor in Florida,* 4–5. The Citizens' Advisory Committee on Migrant and Agricultural Labor held hearings in Belle Glade, Homestead, Immokalee, and Winter Haven in the spring of 1956. The committee included Andrew Duda, of Duda Brothers, one of the largest growers in Florida, and Doyle Carlton, later Florida's secretary of agriculture. Jones, *The Dispossessed,* 178.
5. *Migrant Farm Labor in Florida,* 10.
6. Ibid., 5.
7. Hahamovitch, *Fruits of Their Labor,* 3–7.
8. Ibid., 5.
9. Craig, *Bracero Program,* ix; Jones, *The Dispossessed,* 172.
10. Griffith and Kissam, *Working Poor,* 5.
11. Hahamovitch, *Fruits of Their Labor,* 124–25, 227.
12. Griffith and Kissam, *Working Poor,* 243–46.
13. Romo, interview. A family is said to have settled out when it no longer migrates with the crops but makes a home in one area and works local farms.
14. Schell, interview.
15. National Safety Council, *Report on Injuries in America.*
16. Martin and Martin, *Endless Quest,* 168–69.
17. Romo, interview.

18. NIOSH, *Report on Highest Death Rates.*
19. Mobed, Gold, and Schenker, "Occupational Health Problems," 368.
20. Ibid.

Chapter 1. Moving with the Crops

1. Coles, *Migrants, Sharecroppers, and Mountaineers,* 61–63.
2. Romo, interview.
3. Schell, interview.
4. Goyette, *Farmworker Needs—Agency Services,* 17.
5. Medina, interview.
6. U.S. Bureau of Labor Statistics, analysis of data from 2000, "National Census of Fatal Occupational Injuries."
7. Testimony of David Moody and Gregory S. Schell, Florida Rural Legal Services, before the House Economic Worker Protections Committee for Agricultural Workers, U.S. House of Representatives, May 25, 1995.
8. Goyette, *Farmworker Needs—Agency Services,* 36. Based on indications from the most recent census, though, these statistics are not yet completely compiled. Approximately 20 percent of Florida's labor force either carpooled or walked to work in 1990. United States Census, 1990.
9. Metro Report, Traffic Accidents, *Palm Beach Post,* December 31, 1996.
10. Ibid.
11. "Study of the Migrant Labor Report."
12. Michele Koidin. "Seven Immigrants Found Dead in California Desert," *The Charleston Gazette,* August 14, 1998.
13. "Florida Cash Receipts by Commodity, 1997," *Florida Agricultural Facts* (1997).
14. Ibid.
15. Avocado statistics from Florida Agricultural Statistics Service (FASS) (Citrus), available at http://www.nass.usda.gov/fl/rtoc0v.htm.
16. Romero, interview.
17. Baheña, interview.

Chapter 2. Wages

1. Florida Department of Agriculture, FASS, *Farm Labor Report.*
2. Ibid.
3. Ibid.
4. Deborah O'Neil, "Hunger Strike on Hold," *St. Petersburg Times,* January 19, 1998; Benitez, interview.
5. Mobed, Gold, and Schenker, "Occupational Health Problems," 369.
6. U.S. Dept. of Labor, *A Demographic and Employment Profile of United States Farmworkers,* 37.
7. Asbed, interview; Benitez, interview.

8. Schell, interview.

9. Ibid.

10. Benitez, interview.

11. Asbed, interview.

12. Ibid.

13. Statement of Immokalee workers regarding the boycott of Taco Bell.

14. Germino, interview.

15. Mireya Navarrro, "Group Forced Illegal Aliens into Prostitution, U.S. Says," *New York Times*, April 24, 1998.

16. Eric Brazil, "Strawberry Grower to Pay Back Wages Totaling $575,000," *San Francisco Examiner*, August 28, 1997.

17. Verena Dobnik, "Strawberry Workers March in New York," Associated Press, *New York Times*, March 28, 1998.

18. Gerald Ensley, "Quincy Mushroom Workers File Suit," *Tallahassee Democrat*, March 9, 1998; Susan Salisbury, "A New Harvest of Union Members," *Miami Daily Business Review*, December 10, 1999.

19. Salisbury, "A New Harvest of Union Members."

20. Compiled information from Cuevas Sr. and Schell interviews.

21. *Crispin Calderon, et al., Plaintiffs, v. Jim Witvoet, Sr.*, House Testimony; see http://www.ca.7.uscourts.gov/op

22. Ibid.

23. Cuevas Jr., interview.

24. Hahamovitch, *Fruits of Their Labor*, 202.

25. "Study of the Migrant Labor Report," 13; Ryan, "Coops and Canneries," 81, app. 3.

Chapter 3. Housing

1. Romo, interview.

2. Ibid.

3. Romero, interview.

4. Charlie Whitehead, "Pueblo Bonito Dedicated in Bonita Springs," *Naples Daily News*, June 7, 1999.

5. Ibid.

6. Jorn, interview.

7. Ibid.

8. Bulletin from United Farmworkers, February 2000, available at http://www.ufw.org.

9. Brown and Sizemore, interview.

10. Ibid.

11. Ibid.

12. "A New Day Dawns in Tommytown," editorial, *Tampa Tribune*, November 12, 1999.

13. Public Health General Provisions, Sec. 1–5, 1999.

14. Wayne Washington, "Donated Ticket Will Fly Man to Family's Burial," *St. Petersburg Times*, December 22, 1998.

Chapter 4. Education

1. *Basic Programs Operated by Local Educational Agencies.*
2. Peres, interview.
3. Cannon, interview.
4. Black, interview. Black is director of the adult education facility at Beth-El Mission in Wimauma, which is funded by the Migrant Education Program of Hillsborough County. Students must be at least sixteen years old and have no secondary education degree to be accepted.
5. Romo, interview.
6. Coles, *Migrants, Sharecroppers, and Mountaineers*, 65–66.
7. Griffin, interview.
8. Coles, *Migrants, Sharecroppers, and Mountaineers*, 77.

Chapter 5. Health and Safety

1. Goldfarb, *Migrant Workers*, 34–42.
2. Ryan, "Coops and Canneries," 37.
3. Ibid., 38.
4. Himelick, interview.
5. "Farmworker Women's Health Project."
6. Ibid.
7. Romo, interview; Center for Disease Control, *Prevention and Control of Tuberculosis*, 2.
8. Center for Disease Control, *Prevention and Control of Tuberculosis*, 1.
9. Ibid., 2.
10. S. D. Ciesielski, "TB among North Carolina Migrant Farm Workers," *Journal of the American Medical Association*, 1715–19.
11. Florida Department of Health, *Public Health Statistics.*
12. Ibid.
13. Ibid.
14. Ascencio, interview; Himelick, interview.
15. Title 29 of the *Fair Labor Standards Act*, Subtitle B, CFR. 570.71.
16. National Institute for Occupational Safety and Health (NIOSH), "Childhood Agricultural Injury Prevention Initiative."
17. Ibid., 3
18. Ibid., 4.
19. Title 29 of the *Fair Labor Standards Act*, Subtitle B, CFR. 575.1.
20. Florida Department of Labor.

21. Romo, interview.
22. NIOSH, *Children and Agriculture: Opportunities for Safety and Health,* Agricultural Injury Prevention report, 1.
23. National Safety Council, "Survey of Mortality and Morbidity among Farmworkers," *Report of Injuries in America.*
24. Mobed, Gold, and Schenker, "Occupational Health Problems," 370.
25. Ibid.
26. Ibid., 375.

Chapter 6. Pesticides

1. *Florida Toxic Use Reduction Bill,* Amendment, SB 1208/HB 1031.
2. "Benlate Case Will Go to Jury." *Miami Herald,* June 6, 1996; John Pacenti, "Benlate Jury Awards $4 Million to Family of Eyeless Child," *Detroit News,* June 8, 1996.
3. Ibid.
4. "Benlate Case," *Miami Herald,* June 6, 1996.
5. Rauber, "Growers Seek to Delay Methyl Bromide Ban," 20, 21; California Environmental Protection Agency, *Report on Methyl Bromide Use,* 19; U.S. Environmental Protection Agency, *Methyl Bromide Use.* Nationwide, strawberries are the second largest crop on which methyl bromide is used, with tomatoes retaining the number one ranking. California EPA, *Report on Methyl Bromide Use,* 20.
6. California EPA, *Report on Methyl Bromide Use,* 19–20; EPA, *Methyl Bromide Use.*
7. *Reaping Havoc,* 6–7.
8. Ibid., 7.
9. Ibid., 27.
10. Ibid., 16.
11. Moses, "Examining Environmental Health," 18.
12. Ibid., 19.
13. "Claws out over Delaney Clause," 10.
14. Friends of Lake Apopka website, "History of Lake Apopka."
15. Ibid.; Colburn, Dumanoski, and Myers, *Our Stolen Future,* 8.
16. Connor, interview.
17. Friends of Lake Apopka website, "History of Lake Apopka"; Blake, *Land into Water—Water into Land,* 120–29.
18. Colburn, Dumanoski, and Myers, *Our Stolen Future,* 150–54.
19. Connor, interview.
20. Ibid.
21. Ibid.
22. Economos, interview.
23. Ibid.
24. Connor, interview.
25. Pitter, interview.

26. "Farmworker Women's Health Project."
27. EPA, *Methyl Bromide Use.*
28. Marlene Sokol, "Still a Bitter Harvest," *St. Petersburg Times,* June 25, 1995.
29. EPA, *Methyl Bromide Use.*
30. Solomon, *Trouble on the Farm,* 7.
31. Ibid.
32. Ibid., 2; EPA, *Protecting Children from Pesticides,* January 2002.
33. EPA, *Children and Pesticides,* report, 1991; Amy Ellis, "Plan Takes Work on Pesticides to Fields in Pasco," *St. Petersburg Times,* November 4, 1999, Hernando edition.

Chapter 7. Immigration

1. Larry Dougherty, "Settlement Near in Suit over Death," *St. Petersburg Times,* December 4, 1997.
2. Cuevas Sr., interview; Montweiler, *Immigration Reform Law of 1986,* 7.
3. Dan Herbeck, "INS Accused of Mistreating Workers," *Buffalo News,* November 8, 1997.
4. Montweiler, *Immigration Reform Law of 1986,* 3–5.
5. Hahamovitch, *Fruits of Their Labor,* 10.
6. Rothenberg, *With These Hands.* 218–19.
7. Cuevas Sr., interview.
8. Most advocates use the term *H2* interchangeably with *H2A.* Both refer to workers who are recruited outside the United States, issued temporary work visas for the period they work, and then returned to their homeland.
9. Cuevas Sr., interview.
10. Santolini, *The New Americans,* 137–38.
11. Sister Teresa, interview.
12. Kelleher, interview.
13. Hurston, *Their Eyes Were Watching God,* 279–85.
14. Hahamovitch, *Fruits of Their Labor,* 78.
15. McCally, *The Everglades,* 171–73.
16. Schell, interview.
17. Ibid.

Chapter 8. Family Life

1. Martinez, interview.
2. Olga Ros, interview.
3. Romo, interview.

Bibliography

Basic Programs Operated by Local Educational Agencies, Part A of Chapter 1 of Title I, Elementary and Secondary Education Act of 1965, Policy Manual, amended by the Augustus F. Hawkins–Robert T. Stafford Elementary and Secondary School Improvement Amendments of 1988 (Public Law 100–297). Washington, D.C.: U.S. Department of Education, Office of Elementary and Secondary Education, 1991.

Blake, Nelson M. *Land into Water—Water into Land: A History of Water Management in Florida*. Tallahassee: Florida State University Press, 1980.

California Environmental Protection Agency. *Report on Methyl Bromide Use.* Sacramento, 1998.

Center for Disease Control. *Prevention and Control of Tuberculosis in Migrant Farm Workers.* Recommendations of the Advisory Council for the Elimination of Tuberculosis. June 1992.

"Claws out over Delaney Clause." *Chemistry and Industry* 23, no. 2 (February 20, 1995).

Colburn, Theo, Dianne Dumanoski, and John Peterson Myers. *Our Stolen Future.* New York: Plume Books, 1997.

Coles, Robert. *Migrants, Sharecroppers, and Mountaineers.* Vol. 2 of *Children of Crisis.* Boston: Little, Brown, 1971.

Craig, Richard. *The Bracero Program: Interest Groups and Foreign Policy.* Austin: University of Texas Press, 1971.

Douglas, Marjory Stoneman. *The Everglades: River of Grass.* St. Simons Island, Ga.: Mockingbird Books, 1974.

"Farmworker Women's Health Project." *The Network News* (National Women's Health Network) 17, no. 6 (November 1992): 6.

Florida Department of Agriculture. Florida Agricultural Statistics Service, *Farm Labor Report.* November 23, 1999. http://nass.usda.gov/fl/rtoc0cihtm.

Florida Department of Health and Human Services, *Housing Opportunities for People with AIDS* (HOPWA), 2002. http://www9.myflorida.com/aids/care/hoptwa.html

———. *Public Health Statistics,* Public Health Indicators Data System (P.H.I.D.S.) http://www9.myflorida.com/planning_eval/intro.html

Florida Department of Labor, Department of Business and Professional Regulation, Division of Professions, Farm and Child Labor Program. *Child Labor Overview.* www.state.fl.us/dbpr/pro/farm/compliance/childlabor/overview.shtml

Friends of Lake Apopka. http://www.fola.org.

Goldfarb, Ronald L. *Migrant Workers: A Case of Despair.* Ames: Iowa State University Press, 1981.

Goyette, Cherie, project director. *Farmworker Needs—Agency Services: A Study of Migrant and Seasonal Farmworkers and Service Agencies in a Four County Central Florida Area.* Orlando: Florida Technological University, Office of Graduate Studies and Research, 1977.

Griffith, David, and Ed Kissam. *Working Poor: Farmworkers in the United States.* Philadelphia: Temple University Press, 1995.

Hahamovitch, Cindy. *Fruits of Their Labor: Atlantic Coast Farmworkers and the Making of Migrant Poverty.* Chapel Hill: University of North Carolina Press, 1997.

Howard, Robert West. *The Vanishing Land.* New York: Villard Books, 1985.

Hurston, Zora Neale. *Their Eyes Were Watching God.* New York: Lippincott, 1937.

Hurt, R. Douglas. *American Agriculture: A Brief History.* Ames: Iowa State University Press, 1994.

Jones, Jacqueline. *The Dispossessed: America's Underclasses from the Civil War to the Present.* New York: Basic Books, 1992.

Martin, Philip L., and David A. Martin. *The Endless Quest: Helping America's Farm Workers.* San Francisco: Westview Press, 1994.

McCally, David. *The Everglades: An Environmental History.* Gainesville: University Press of Florida, 1999.

Migrant Farm Labor in Florida: A Summary of Recent Studies. State of Florida Legislative Council and Legislative Reference Bureau. Commissioned by Governor LeRoy Collins. Tallahassee, 1961.

Mobed, Ketty, Ellen B. Gold, and Marc B. Schenker. "Occupational Health Problems among Migrant and Seasonal Farm Workers: Cross-cultural Medicine a Decade Later." *Western Journal of Medicine* 157, no. 3 (September 1992).

Montweiler, Nancy Humel. *The Immigration Reform Law of 1986: Analysis, Text, and Legislative History.* Washington, D.C.: Bureau of National Affairs, Inc., 1987.

Moses, Marion, M.D. "Examining Environmental Health." *Environmental Action Magazine* 6, no. 3 (1994).

National Agricultural Statistics Service (NASS). *1998 Childhood Agricultural Injuries.* Washington, D.C., 1998.

National Institute for Occupational Safety and Health (NIOSH). "Childhood Agricultural Injury Prevention Initiative." July 1999. Available at http://www.niosh.com.

———. *Children and Agriculture: Opportunities for Safety and Health,* Agricultural Injury Prevention report, July 1999.

———. *Report on Highest Death Rates by Industry and Most Deadly Occupations.* Analysis of data from 1990–92. August 1998. Available at http://www.niosh.com.

National Safety Council. *Report on Injuries in America.* 1999.

Rauber, Paul. "Growers Seek to Delay Methyl Bromide Ban." *Sierra Club* 81, no. 4 (July 1996).

Reaping Havoc: The True Cost of Using Methyl Bromide on Florida's Tomatoes. A report by Friends of the Earth; Farmworker Association of Florida; Farmworkers Self-Help, Inc.; Florida Consumer Action Network; and Legal Environmental Assistance Foundation. Washington, D.C.: Friends of the Earth, 1998.

Rothenberg, Daniel. *With These Hands: The Hidden World of Migrant Farmworkers Today.* Harcourt, Brace, 1998.

Ryan, Susan Eisenberger. "Coops and Canneries: An Analysis of the Differential Success of Two Development Projects for Farmworkers in South Central Florida." Master's thesis, University of South Florida, 1986.

Santolini, Al. *The New Americans: An Oral History, Immigrants and Refugees in the U.S. Today.* New York: Viking Press, 1988.

Solomon, Gina M., M.D. *Trouble on the Farm: Growing Up with Pesticides in Agricultural Communities.* Natural Resources Defense Council report. 1998.

"A Study of the Migrant Labor Report, 'Harvest of Shame,' Presented over the Columbia Broadcasting System Television Network on November 25, 1960. Prepared by a Committee Composed of Local Interested Citizens of Belle Glade, Florida." February 15, 1961. Document collection, University of Florida, Gainesville.

U.S. Bureau of Labor Statistics. "National Census of Fatal Occupational Injuries." Analysis of data from 2000. Available at http://www.bls.gov/news.release/cfoi.toc.htm

U.S. Department of Labor. *A Demographic and Employment Profile of United States Farmworkers: Findings from the National Agricultural Workers Survey, U.S. Department of Labor, 1997–1998.* Research Report 8, March 2000. www.dol.gov/asp/programs/agworker/report_8.pdf

———. *Fair Labor Standards Act,* Wage and Hour Division, Employment Standards Administration, 29 CFR, parts 570, 579. Child Labor Regulations, Orders and Statements of Interpretation Child Labor Violations—Civil Money Penalties; Proposed Rules. http://www.dol.gov/esa/regs/fedreg/proposed/99030776.htm

U.S. Environmental Protection Agency (EPA). *Methyl Bromide Use.* Toxic chemical report. Washington, D.C., 1997. Available at http://www.epa.gov.pesticides/citizens/kidspesticide.htm

———. *Protecting Children from Pesticides*. Office of Pesticide Programs, January 2002. www.epa.gov/scipoly/child

Voegler, Ingolf. *The Myth of the Family Farm: Agribusiness Dominance of U.S. Agriculture*. Boulder, Colo.: Westview Press, 1981.

Interviews

Asbed, Greg. Coordinator, Coalition of Immokalee Workers, Immokalee, Fla.

Ascencio, Caridad. Founder and worker, Caridad Health Care Clinic, Palm Beach County.

Baheña, Alfredo. Pesticide coordinator for Farmworker Association of Florida, Pierson, Fla.

Benetiz, Lucas. Farmworker and activist, Coalition of Immokalee Workers, Immokalee, Fla.

Black, Angie. Director, Migrant Adult Education Program, Beth-El Mission, Wimauma, Fla.

Brown, Marvin, and Jay Sizemore. Strawberry farmers and owners of JayMar Farms, Balm, Fla.

Cannon, Juanita. Assistant director, Migrant Education Program, Hillsborough County.

Connor, Jim. Former Lake Apopka Restoration Project manager. Manager, St. Johns River Water Management District.

Cuevas, Fernando Jr. Florida representative of Farm Labor Organizing Committee (FLOC).

Cuevas, Fernando Sr. National vice president, Farm Labor Organizing Committee (FLOC).

Economos, Jeannie. Former coordinator of Lake Apopka project, Farmworker Association of Florida.

Germino, Laura. Community specialist, Florida Legal Services.

Griffin, Ninfa. Title I ESOL teacher, Hillsborough County.

Himelick, Tom. Associate director, Emory University Physicians Assistant Program, Atlanta, Ga.

Jorn, Evan. Director, Beth-El Mission, Wimauma, Fla.

Kelleher, Maureen. Immigration lawyer, Immigration Advocacy Center, Immokalee, Fla.

Martinez, Mary L. Education coordinator and assistant director, Migrant Head Start, Camp Hacienda, Naples, Fla.

Medina, Sylvia. Farmworker and crew boss, Wimauma, Fla.

Peres, Luana. Director, Migrant Education Program, Hillsborough County.

Pitter, Margie Lee. Displaced Lake Apopka farmworker.

Romero, Ignacio. Director of Immokalee Office, Farmworker Association of Florida.

Romo, Margarita. Director and founder, Farmworkers Self-Help, Dade City, Fla.

Ros, Olga. Wife of Pastor Ramiro Ros of Beth-El Mission, Wimauma, Fla.

Ros, Ramiro. Pastor of Beth-El Mission, Wimauma, Fla.

Schell, Greg. Lawyer, Migrant Farmworker Justice Project, Belle Glade, Fla.

Sister Teresa. Coordinator of Los InDios, a Guatemalan self-help program, Indiantown, Fla.

Index

Italic page numbers denote photographs.

African Americans: compared to Mexican workers, 7, 155; as early migrants, 7–10; as field-workers in Belle Glade, 5–7; northern migration of, 10; recruitment of, as cane cutters, 169; in World War II, 8
Agriculture in Florida: corporate farming, 24; post–Civil War, 7; winter growing season, 8; during World War II, 138;
AIDS, 111–14; outreach clinic for, *115*; single men and, 116
Asbed, Greg, 58, 60, 63.
Ascencio, Caridad, 114–18
avocados, 40

Bahamanians, 7, 10
Bahena, Alfredo, *106, 141*
Balm, Florida, 25, 148
Balm-Wimauma Affordable Housing Project, 79–82
Beans: crates in fields, *17, 35, 50, 154, 168*
Belle Glade, xiv; citizens' report, 4–5, 36, 71; housing in, 163, 172; loading dock, *165*; population of, 163, 171; sign at city limits, *164*; sugar industry and, 169; as winter vegetable capital, 169; Zora Neale Hurston and, 166, 169

Benetiz, Lucas, 55–58; Coalition of Immokalee Workers and, *54, 59*
Benlate: and birth defects, 129; and Castillo family lawsuit, 129, 130
Beth-El Mission, 79, 82; Christmas giveaway, *185–86*; education programs at, 91–94; family services and, 80; students at, *182*; Sunday service at, *183*; women and crafts at, 187; women studying English at, *90*
Big Sugar, xiv, 169
Black, Angie, 91–94, 196n.4
bracero program, 8, 66–67, 155
Brown, Marvin, 82–85, 191
Brownsville, Texas, 25, 156–57
Bush, George W., 175
Bush, Jeb, 60, 96

cabbage, 28, *61*
California: and migrant deaths, 39; and western migrant stream, 8
Camp Hacienda, and Head Start, 179–80
Cancer, incidence in younger women, 109
Cannon, Juanita, 89–91
Caribbean workers: Bahamians, 7; Haitians, 11; Jamaicans, 7, 11; Puerto Ricans, 7
Caridad Health Care Clinic, 114–18

Carter, Jimmy, 48, 55
Center for Disease Control (CDC), 111–13
Chavez, Cesar, 13, 64
Children: early responsibilities of, 22; in the fields, *98*, *101*, *120*, *128*; injury statistics of, 119; NRDC report on pesticide risks to, 148–49; and pesticide exposure, high risk of, 121–22, *128*, 134, 148–50 (*see also* methyl bromide); psychological impact of traveling on, 20–24; safety programs for, 118–19; social education of, 102–4, vaccination programs for, 114; work regulations for, 118–21
Children of Crisis, 22
Cinco de Mayo, 19
citrus: Florida revenue from, 40; Guatemalans and, 159; hazards of picking, *41*; workers picking, *70*
Coalition of Immokalee Workers: and Greg Asbed, 58–63; and Lucas Benetiz, 55–58; ethnic composition of, 48, 58; hunger strike by, 48, 51; and pinhooking, 60; and Taco Bell, 60, 195n.13; and wage demands, 48, 60
Coles, Robert, 22
Collins, Gov. Leroy, 5–7
Connor, Jim, 138, 140, 143
Corporate farms, and preference for single male migrant workers, 171–72
coyotes, 11, 24; immigrant deaths associated with, 39; and slavery, 57–58, 63–64
crew bosses, 20, 24; debt peonage and, 57; farmworker dependence on, 36; fines for transporting illegal workers, 153; physical abuse and, 157; and timekeeping, 25; and transportation, 32–39; unscrupulousness of, 34
cucumbers, picking of, *26*, *27*, *28*, *31*
Cuevas, Fernando, Jr., 69
Cuevas, Fernando, Sr., 69, 156–58

Dade City, 72
Delaney Clause, 134, 137
Dow Chemical, 129
DuPont Chemical Company, 129

Dust Bowl, 7

Economos, Jeannie, 143–44
Education: at Beth-El Mission, *90*, 91–94; effects of migration on, 16; eligibility for aid, 92; government programs, 16, 88–91; language barriers and, 93; Migrant Education Program, 88–91; Migrant Head Start, 178–79
Elementary and Secondary Education Act, and Title I, 88
Emory University, 107
Environmental Protection Agency (EPA), 134, 137, 144; and pesticides, 149
Everglades, the, 11
Everglades agricultural area, 5

Fair Labor Standards Act, 71, 118
families: and celebrations, 188; extended, 22; working together, 25–28
Farm Labor Contractor Registration Act (FLCRA), 13
Farm Labor Organizing Committee: and H2-A workers, 69; and Fernando Cuevas, Sr., 156
Farm Security Administration, and bracero housing, 155
Farmworker Association of Florida: and Immokalee, 75–77; and Lake Apopka, 143–44
Farmworkers: and abusive crew bosses, 157, 169; and affordable housing projects, 75–77, 79, 83–85; African Americans as, 5–8, 10 (*see also* African Americans); average hours worked by, in Florida, 51; in Belle Glade, 4–10; corporate farms and, 24, 171; and debt peonage, 57; dependence of, on crew bosses, 35–36; domestic violence and, 94; effects of weather on, 28, 32; eligibility of, for social programs, 89; and extended families, 19; and field preparation, *38*; in Great Depression, 8; Mexicans as, 5, 7, 8, 10–11, 19; national population statistics of, 16; Native Americans as, 10; in 19th century, 7–8; in 1950s, 5, 10–11; as "offshore workers," 8, 24;

204 Index

percent of single male workers among, 24; population estimates for, 13; recruitment of, 34; rights violations of, 67; single unaccompanied males among, 24, 171; slavery and, 57–58, 63–64; and Social Security, 69, 71, 157; social programs for, 16–19, 44; summary of problems of, 13–19; traffic-related fatalities among, 32–39; white laborers and preferential treatment of, 10; during World War II, 8. *See also* education; housing; immigration; pesticides; wages
—health of: field sanitation for, 107–9; fungal infections among, 109, *110*; musculoskeletal problems of, 107–8, *112*; outreach clinics for, 111, 114–16, *115*, 118; and pesticide poisonings, 116, 130, 147; prevalent diseases among, 111–14; protective clothing for, 28, *65*, *136*; repetitive stress injuries among, 44; requirements of H2 workers, 176; Worker's Compensation and, 158

Farmworkers Self-Help, 75, *95*; health care and, 105; quilting group of, 187; services provided by, *152*; Teen Dream Team and, 96–98

ferns: bundling of, *43*; cultivation of, 44; cutting of, *15*, *142*

field-workers. *See* farmworkers

Florida, as right-to-work state, 13, 52

Florida crops: by county, 46; extended growing season for, 8; major, 40; in winter, 28. *See also* specialty crops

Florida Department of Agriculture, and statistics on Florida farmworkers, 48–51

Florida Department of Health, 85–87

Florida Department of Labor, 69

Florida Farm Bureau, 133

Florida Fruit and Vegetable Association, 51, 133, 157

Florida Rural Legal Services, 63

Food Quality Protection Act of 1996, 149

Fox, Vicente, 175

Friends of the Earth, 130

Gadsden County, 134

Gargiulo, Inc., 55, 57; and California farmworkers, 64; and worker housing, 180

Germino, Laura, 63–64

Graham, Sen. Bob, 60; and proposed farmworker amnesty, 162

Griffin, Ninfa, 100–104

Guatemalans, 10, 47–48; and different dialects, 102, 159–60; and Indiantown, 159–61

H-2 (H2-A) program, 11, 66–67, 198n.8; benefits of, to growers, 156; housing requirements for, 175; proposed foreign worker compromise bill and, 175–76; regulations of, 156; as replacement for bracero program, 155; as strawberry workers, 172; and wage guarantees, 69, 170

Haitian Refugee Act (HRIFA), and amnesty, 161

Haitians: in Belle Glade, *9*, *12*; denial of political refugee status to, 160; in Immokalee, 47–48; picking beans, *154*, *168*

Harvest of Shame, xiv, 4–5, 163, 192

Health. *See* Farmworkers, health of

Hillsborough County: housing in, 1, 79, 82–85; job training and education programs in, 91–93; and Migrant Education Program, 88–91; workers poisoned in, 148

Himelick, Tom, 107–9

HIV. *See* AIDS

Homestead, Florida, 32

House Ecomonic Worker Protection Committee, 34

Housing: in Belle Glade, 163, 166, *167*; and cane cutters, 169; code violations and, 79, 82; farmworker's house, *73*; government allocations for, 77, 85; in Hillsborough County, 79; in Immokalee, 75–77; kitchens in, *78*, *84*; in labor camps, *81*, *106*; laundry facilities in, *74*, *86*; overcrowding in, 102; and Pueblo Bonito project, 77; rental costs of, 179–80; sanitation and, *106*, *117*; for single men, 24, 171–72; temporary, 20–22; trailers as, *3*, *76*, *80*

Housing Opportunities for People with AIDS (HOPWA), 114

Huerta, Dolores, 64–65

Hurston, Zora Neale, 166, 169

immigrants: as farmworkers in 19th century, 7
—illegal, 24; deaths of, 39, 151; harassment of, by authorities, 151, 153; lack of advocacy for, 172
immigration: and green cards, 28; quota system and reforms, 153
Immigration and Nationality Act of 1952, revision of, 153
Immigration and Naturalization Service (INS), 39; and apprehending undocumented workers, 153
Immigration Reform and Control Act of 1986, 16, 69, 151; and elimination of quota system, 153; as provision for accessing Third World labor, 156
Immokalee, 37; and Haitians, 161–62; housing in, 75–77. See also Coalition of Immokalee Workers
Interstate Commerce Commission, 36

Jamaicans, 7; as H-2A workers, 11
Jamestown, 7
Jorn, Evan, 79, 82, 92, 180

Kanjobal Indians. See Guatemalans
Kelleher, Maureen, 161–62
kumquats, 42

labor camps. See housing
Lake Apopka, 137–47; alligators, underdeveloped in, 138, 140; chemical runoff and, 138; drainage of, for muck farming, 138, 140; Jeannie Economos and, 143–45; and federal farmland buyout program for restoration, 140, 143–44; labor camp, 81; wildlife deaths in, 137, 143–44; workers displaced by restoration of, 143–47; Zellwood Drainage and Water Control District and, 140
Lake Okeechobee, 4, 8–11; and Belle Glade, 166–69; and sugarcane farming, 163

mangoes, 40
Martinez, Mary, 179–80

Mayans, 11, 102
Medina, Narcisso, 25
Medina, Sylvia, 25–32, 26
Methyl bromide, 130, 132–34, 197n.5; and exposure of children to, 134; and Montreal Protocol, 133
mevinphos. See pesticides
Mexican labor program. See bracero program
Mexicans: as braceros, 8; and extended families, 19; relation to homeland, 11; unaccompanied male workers, 24; undocumented workers, 11, 24. See also Farmworkers
Migrant Education Program, 88–91
Migrant Farm Labor in Florida: A Summary of Recent Studies, 5, 193n.4
Migrant Farmworker Justice Project, 35; in Belle Glade, 166; and undocumented workers, 52, 55
Migrant Head Start, 178
Migrant and Seasonal Agricultural Worker Protection Act (AWPA) 67
migrant streams, 8–9. See also migration
migrant workers, definition of, 5. See also seasonal worker
migration: in depression-era Florida, 8; patterns of, 22
Monsanto, 129
Moses, Marion, 134
muck farming. See Lake Apopka; Lake Okeechobee
Murrow, Edward R., xiv, 4–5, 19, 163, 192
mushrooms, 66

National Committee for Childhood Agricultural Injury Prevention (NCCAIP), 119
National Farmworkers Association. See United Farm Workers
National Institute for Occupational Safety and Health (NIOSH): and work-related deaths, 32
National Resources Defense Council (NRDC), 148–49
National Safety Council: farm injuries, study of, 122–25

Native Americans: Creeks and Cherokees as farmworkers, 10; Guatemalans and Mann dialect, 102; Kanjobal Indians as Mayan descendants, 159
New Jersey, and truck farms, 7
Nicaraguan Adjustment and Relief Act (NICARA), 160
North Carolina, as importer of H2-A workers, 157–58

Occupational Safety and Health Administration (OSHA): and field sanitation, 107; and injury report regulations, 125
Office of Migrant Health, Dept. of Health and Human Services, 111
offshore workers. *See* Caribbean workers; H-2 program
O'Loughlin, Father Frank, 159
"Operation Gatekeeper," 39
packinghouses, 10, *30*
padrone. *See* crew boss
Palm Beach County, migrant population in 1955, 4–5
parathion. *See* pesticides
paraquat. *See* pesticides
Pasco County, 72, 85; and pesticide safety campaign, 149
PELL grants, 92
peppers, 28; sorting, *30*
Peres, Luana, 89–91
pesticides: Benlate, 129–30; certification for handling of, 126; and Delaney Clause, 134, 137; developed as chemical weapons in World War II, 134, 147; discarded containers of, *127*; effects on environment, 130 (*see also* Lake Apopka); exposure of children to, in fields, *128*, 148–50; in field preparation, 40; language as barrier to safe pesticide handling, 150, 158; methyl bromide, 130–34; mevinphos, 148; paraquat as herbicide, 139; parathion, 134; posted regulations for use of, *131*; residues in farmworkers' homes, 149; and "Right to Know" bill, 126; sprayed in crop dusters, 134, 147; symptoms of exposure to, 116, 126, 147; used in field preparation, *38*
Phosdrin. *See* pesticides: mevinphos
piecework pay rate. *See* wages
Pierson, Florida, 44; fern farms in, *141*; labor camp in, *106*
Pitter, Margie Lee, 145–47, *146*
political refugees: from Central America and southern Mexico, 10; Guatemalans as, 160
Pueblo Bonito. *See* housing
Puerto Ricans, 7

Quincy Farms: and mushroom workers' protest, 52; workers' settlement, 66, 190

Romero, Ignacio, 75–77
Romo, Margarita: and Farmworkers Self-Help, 75, 94–99; and Teen Dream Team, 96–99, *97*
Roosevelt, Eleanor, 172
Ros, Olga, 92, 184
Ros, Father Ramiro, 92, *183*, 184

Saffold Farms, 25, 32
sarin, 147
saw palmetto berries, 40, 63
Schell, Greg, *12*, 34, 52, 55, 170–76, *173*
settle out, 11, 22–32; benefits of, 24, 193n.13
sharecroppers, 10
sindocumentos. *See* illegal workers
Sister Teresa, 159–61
Sizemore, Jay, 82–85, 191
South Carolina, 25, 28
specialty crops, 40–44
Steinem, Gloria, 64
St. Johns River Water Management District, 138, 140. *See also* Lake Apopka
strawberries: and California reforms, 65–66; picking of, *14*, 64, *65*
Stuart, Mike, 51
sugar industry, 10; government subsidies for, 169; and H-2A workers, 11, 155–56, 169, 175–76; and mechanical cane harvesting, 11, 13, 24, 155, 170

Index 207

sweet potatoes, 25, *45*

Taco Bell. *See* Coalition of Immokalee Workers
tenant farmers, 10
Texas: Brownsville, 25, 156; and Texas Mexicans, 5–19
Their Eyes Were Watching God, 166, 169
Title I. *See* Elementary and Secondary Education Act
tobacco: in Florida, 25; in Jamestown, 7
tomatoes: picking of, 25, *29, 49, 56, 120*; revenue from Florida crop, 40
trailers, 3; and Immokalee housing project, 75–77, *76*; in isolated locations, *80*; and kitchens, *78*
transportation: buses taking workers to fields, *2, 33, 37, 68, 165*; unreliable vehicles, 32–34; vehicle ownership, 3
truck farms, 7
tuberculosis, 111–13

United Farm Workers (UFW), 13, 52, 82; and AFL/CIO affiliation, 13, 64; Cesar Chaves and founding of, 13; and Quincy Farms mushroom workers, 52, 66
Uribe, Ignacio, and family, *21, 23*
U.S. Department of Education: and Office of Migrant Education and Migrant Student Record Transfer System, 16
U.S. Department of Immigration: and reforms of 1996, 67
U.S. Sugar Corporation, 155, 170

vehicles: ownership of, 34; reliability of, 32–34, *36*
Virginia, 7, 25
Volusia County, 43

Wages: H2 workers and, 171; inflated estimates by growers, 52; lack of inflationary adjustment of, 52; national average for farmworkers, 51; for piecework, 48–49, 158; required minimum wage, 48; unemployment insurance and, 171
Wimauma, Florida, 25; and affordable housing project, 79
Women, health of: birth defects of children and, 129; and domestic abuse, 187; and migrant slavery, 58; miscarriages by, 145–47; and work-related disease, 109
Worker Protection Act of 1995, 158
World War II: and bracero program, 8, 155; and fear of food shortages, 138, 155; and nerve gas development, 134, 147

Zarate, Ramona, selling tacos, *53*
Zucchini, packing of, *62*

Nano Riley is a native Floridian and freelance writer from St. Petersburg. She has written for publications nationwide for twenty-five years. As a reporter for the *St. Petersburg Times* for almost ten years, she wrote many articles on Florida and continues to write about social and environmental issues involving her home state.

Davida Johns has been photographing for social change since the early 1990s. She has traveled through Central America and spent two years in Belize as a Peace Corps volunteer.

The Florida History and Culture Series
Edited by Raymond Arsenault and Gary R. Mormino

Al Burt's Florida: Snowbirds, Sand Castles, and Self-Rising Crackers, by Al Burt (1997)

Black Miami in the Twentieth Century, by Marvin Dunn (1997)

Gladesmen: Gator Hunters, Moonshiners, and Skiffers, by Glen Simmons and Laura Ogden (1998)

"Come to My Sunland": Letters of Julia Daniels Moseley from the Florida Frontier, 1882–1886, by Julia Winifred Moseley and Betty Powers Crislip (1998)

The Enduring Seminoles: From Alligator Wrestling to Ecotourism, by Patsy West (1998)

Government in the Sunshine State: Florida since Statehood, by David R. Colburn and Lance deHaven-Smith (1999)

The Everglades: An Environmental History, by David McCally (1999), first paperback edition, 2001

Beechers, Stowes, and Yankee Strangers: The Transformation of Florida, by John T. Foster Jr. and Sarah Whitmer Foster (1999)

The Tropic of Cracker, by Al Burt (1999)

Balancing Evils Judiciously: The Proslavery Writings of Zephaniah Kingsley, edited and annotated by Daniel W. Stowell (1999)

Hitler's Soldiers in the Sunshine State: German POWs in Florida, by Robert D. Billinger Jr. (2000)

Cassadaga: The South's Oldest Spiritualist Community, edited by John J. Guthrie, Phillip Charles Lucas, and Gary Monroe (2000)

Claude Pepper and Ed Ball: Politics, Purpose, and Power, by Tracy E. Danese (2000)

Pensacola during the Civil War: A Thorn in the Side of the Confederacy, by George F. Pearce (2000)

Castles in the Sand: The Life and Times of Carl Graham Fisher, by Mark S. Foster (2000)

Miami, U.S.A., by Helen Muir (2000)

Politics and Growth in Twentieth-Century Tampa, by Robert Kerstein (2001)

The Invisible Empire: The Ku Klux Klan in Florida, by Michael Newton (2001)

The Wide Brim: Early Poems and Ponderings of Marjory Stoneman Douglas, edited by Jack E. Davis (2002)

The Architecture of Leisure: The Florida Resort Hotels of Henry Flagler and Henry Plant, by Susan R. Braden (2002)

Florida's Space Coast: The Impact of NASA on the Sunshine State, by William Barnaby Faherty, S.J. (2002)

In the Eye of Hurricane Andrew, by Eugene F. Provenzo Jr. and Asterie Baker Provenzo (2002)

Florida's Farmworkers in the Twenty-first Century, text by Nano Riley, photographs by Davida Johns (2003)